WALCH EDUCATION

Daily Warm-Ups ALGEBRA

Common Core State Standards

Betsy Berry, Ph.D.
Indiana University–Purdue University Fort Wayne

1 2 3 4 5 6 7 8 9 10
ISBN 978-0-8251-6883-3

J. Weston Walch, Publisher
40 Walch Drive • Portland, ME 04103
www.walch.com
Printed in the United States of America

Table of Contents

Introduction

Daily Warm-Ups: Algebra, Common Core State Standards is organized into four sections, composed of the high school conceptual categories designated by the Common Core State Standards Initiative: Number and Quantity; Algebra; Functions; and Statistics and Probability. Each warm-up addresses one or more of the standards within these domains.

The Common Core Mathematical Practices standards are another focus of the warm-ups. All the problems require students to "make sense of problems and persevere in solving them," "reason abstractly and quantitatively," and "attend to precision." Many of the warm-ups ask students to "look for and express regularity in repeated reasoning" when generating functions to symbolize patterns or rules. Students must "look for and make use of structure" when factoring and reorganizing expressions and equations to elicit the properties of the function being modeled. Several problems require the use of a graphing calculator; for other problems, graphing calculators make calculations more efficient and allow students to maintain their focus to "construct viable arguments" and prove or disprove conjectures. This provides students the opportunity to "use appropriate tools strategically." A full description of these standards can be found at www.corestandards.org/the-standards/mathematics/introduction/standards-for-mathematical-practice/.

The warm-ups are organized by domains rather than by level of difficulty. Use your judgment to select appropriate problems for your students.* The problems are not meant to be completed in consecutive order—some are stand-alone, some can launch a topic, some can be used as journal prompts, and some refresh students' skills and concepts. All are meant to enhance and complement high school algebra programs. They do so by providing resources for teachers for those short, 5-to-15-minute interims when class time might otherwise go unused.

***You may select warm-ups based on particular standards using the Standards Correlations document on the accompanying CD.**

About the CD-ROM

Daily Warm-Ups: Algebra, Common Core State Standards is provided in two convenient formats: an easy-to-use, reproducible book and a ready-to-print PDF on a companion CD-ROM. You can photocopy or print activities as needed, or project them on a large screen via your computer.

The depth and breadth of the collection give you the opportunity to choose specific skills and concepts that correspond to your curriculum and instruction. The activities address the following Common Core State Standards for high school mathematics: Number and Quantity; Algebra; Functions; and Statistics and Probability. Use the table of contents, the title pages, and the standards correlations provided on the CD-ROM to help you select appropriate tasks.

Suggestions for use:

- Choose an activity to project or print out and assign.
- Select a series of activities. Print the selection to create practice packets for learners who need help with specific skills or concepts.

Part 1: Number and Quantity

Overview

The Real Number System

- Extend the properties of exponents to rational exponents.

Vector and Matrix Quantities

- Perform operations on matrices and use matrices in applications.

Exponential Expressions

The following exponential expressions can be rewritten in other forms. Use your understanding of how exponents behave to change each expression to an equivalent one.

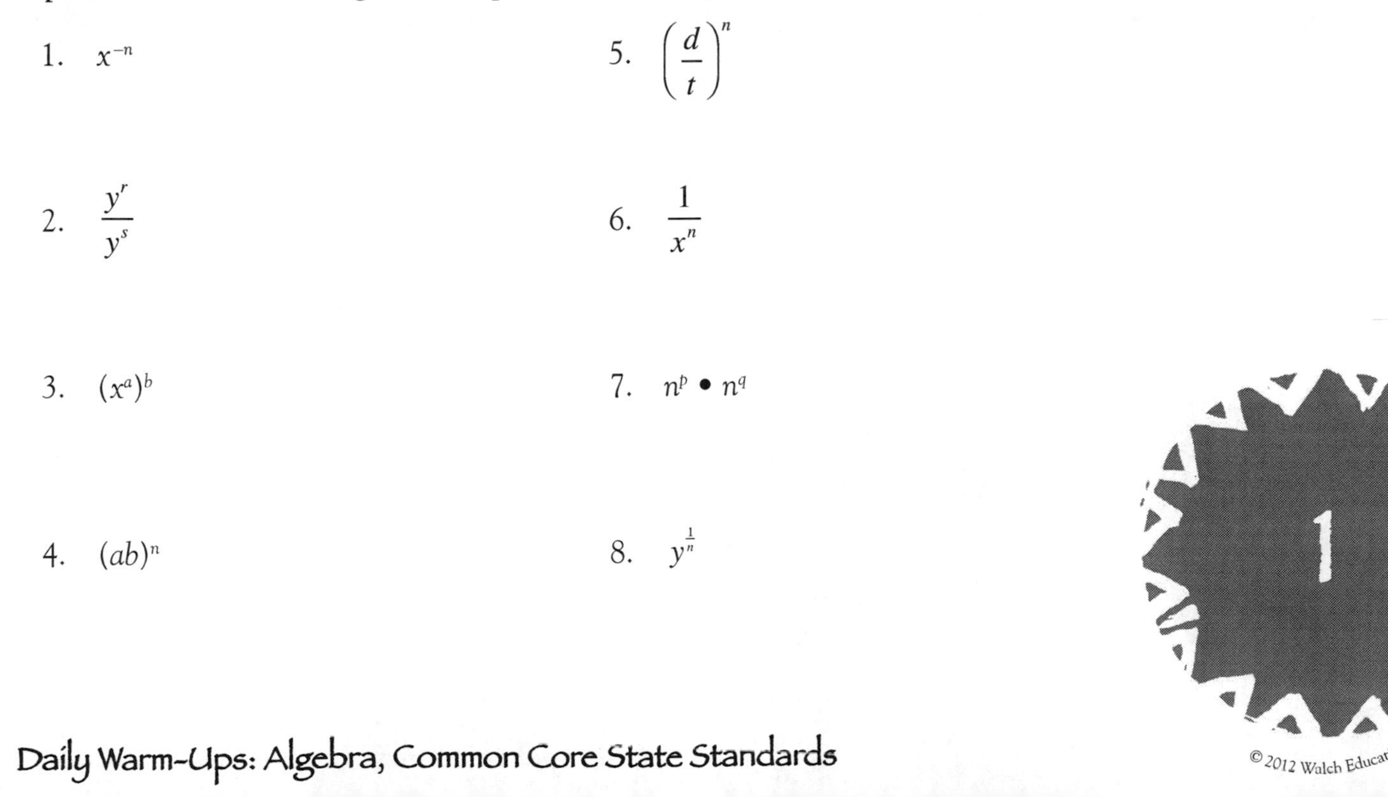

1. x^{-n}
2. $\dfrac{y^r}{y^s}$
3. $(x^a)^b$
4. $(ab)^n$
5. $\left(\dfrac{d}{t}\right)^n$
6. $\dfrac{1}{x^n}$
7. $n^p \bullet n^q$
8. $y^{\frac{1}{n}}$

Intercepting Intercepts

Do all quadratic functions have real-number x-intercepts? Do they all have real-number y-intercepts? Justify your answers to these questions by generating sample functions symbolically and/or graphically. Then explain if and why the functions have real-number intercepts and what the graphs will look like.

Thinking About Using Matrices

Think about a real-world context for creating matrices. What are some advantages and disadvantages of organizing and representing the information in a matrix format? Explain how data might be represented in matrices in more than one way. Write a few sentences to explain your thinking.

Bicycle Sales Matrices

The first matrix below represents the combined quarterly sales of three brands of mountain bikes and hybrid bikes at Bikes Unlimited in three large American cities. The second matrix shows the quarterly sales of hybrid bikes.

Combined Sales

	Trek	K2	Schwinn
Los Angeles	150	95	105
Boston	75	75	90
Chicago	85	110	175

Hybrid Sales

	Trek	K2	Schwinn
Los Angeles	50	40	70
Boston	30	35	50
Chicago	55	50	150

Find the matrix that represents the quarterly sales of mountain bikes. Find the matrix that represents the quarterly sales of mountain bikes and hybrid bikes in Boston for the three brands. Then construct one matrix that shows the total sales for all three brands in all three cities combined. Label the rows of your matrix with the brand names. Label the columns with the terms *hybrid* or *mountain*. What matrix operations did you use to construct each of these matrices?

Manipulating Matrices

Perform the indicated operations on the given matrices and scalars without using a calculator.

$s_1 = 4,\ s_2 = -3$

$[A] = \begin{bmatrix} 3 & 6 \\ 4 & 2 \end{bmatrix}$

$[B] = \begin{bmatrix} -2 & 4 \\ 5 & -1 \end{bmatrix}$

$[C] = [-2 \quad 4 \quad 3]$

$[D] = [3 \quad -2]$

$[E] = \begin{bmatrix} -1 & -5 \\ 3 & -4 \\ 2 & 3 \end{bmatrix}$

$[F] = \begin{bmatrix} 3 & -2 & 4 \\ -1 & 5 & -3 \end{bmatrix}$

1. $[A] + [B]$
2. $s_1 \bullet [B]$
3. $[B] - [A]$
4. $s_2 \bullet [C]$
5. $[F] - [E]$

More Manipulating Matrices

Perform the indicated operations on the given matrices and scalars without using a calculator.

$s_1 = 4, \; s_2 = -3$ $\quad [A] = \begin{bmatrix} 3 & 6 \\ 4 & 2 \end{bmatrix}$ $\quad [B] = \begin{bmatrix} -2 & 4 \\ 5 & -1 \end{bmatrix}$ $\quad [C] = \begin{bmatrix} -2 & 4 & 3 \end{bmatrix}$

$[D] = \begin{bmatrix} 3 & -2 \end{bmatrix}$ $\quad [E] = \begin{bmatrix} -1 & -5 \\ 3 & -4 \\ 2 & 3 \end{bmatrix}$ $\quad [F] = \begin{bmatrix} 3 & -2 & 4 \\ -1 & 5 & -3 \end{bmatrix}$

1. $s_1 \bullet [A] - s_2[B]$
2. $[A] - [B]$
3. $[B] + [D]$
4. $[E] \bullet [F]$

Thinking About Multiplying Matrices

Think about the process of multiplying matrices. Describe in detail how to multiply two matrices and any limitations there might be. Does the order that you multiply matrices matter? Write a few sentences to explain your thinking.

Thinking About Matrix Operations

Think about and create some sample matrices. Describe the conditions when matrices can be added and multiplied. What are the dimensions of the sums and products of matrices? Define and give examples of an inverse of a matrix, an opposite of a matrix, an identity matrix, and a matrix that will not have an inverse.

Part 2: Algebra

Overview

Seeing Structure in Expressions

- Interpret the structure of expressions.
- Write expressions in equivalent forms to solve problems.

Arithmetic with Polynomials and Rational Expressions

- Rewrite rational expressions.

Creating Equations

- Create equations that describe numbers or relationships.

Reasoning with Equations and Inequalities

- Understand solving equations as a process of reasoning and explain the reasoning.
- Solve equations and inequalities in one variable.
- Solve systems of equations.
- Represent and solve equations and inequalities graphically.

Refreshing Radical Ideas

In your work with algebraic expressions and equations, you will encounter real and complex numbers of many forms. Sometimes it is helpful to rewrite them in equivalent forms. Convert each expression below into an equivalent form in which the denominator contains a simple rational term, and all possible like terms have been combined.

1. $\dfrac{6}{\sqrt{3}}$

2. $\dfrac{2}{\sqrt{6}}$

3. $\dfrac{\sqrt{15}}{\sqrt{3}}$

4. $3\sqrt{2}+2\sqrt{32}$

5. $\dfrac{21\sqrt{150}}{7\sqrt{2}}$

Quincy's Quiz

Quincy has been confused about simplifying algebraic expressions. His brother Eli prepared this quiz for him to practice. Look at Quincy's responses below. If the response is correct, write *correct*. If it is incorrect, write the correct answer.

	Expression	Quincy's response
1.	$4y - 5y$	$-y$
2.	$3x + 2x$	$5x^2$
3.	$2n^2 + 4n^2$	$6n^4$
4.	$2x \bullet 3y$	$5xy$
5.	$2a - (a - b)$	$a - b$
6.	$\frac{n+5}{n}$	5
7.	$4 - (n + 7)$	$-3 - n$
8.	$\frac{2x^6}{x^3}$	$2x^2$
9.	$(2xy)^2$	$4x^2y^2$
10.	$2a + 3b + 3a + 4b$	$6a^2 + 12b^2$

Dancing to the Beat of Logarithms

Use symbolic reasoning and the properties of exponents and logarithms to simplify the expression or solve the equation for x.

1. $5 = 2(1.3)^x$

2. $60 = 5\left(1 + \frac{.025}{12}\right)^x$

3. $4\log 8 + 3\log 5 - 3\log 2 = ?$

4. $\log x^3 + \log x^2 + 3\log x = ?$

5. $2^{2x+2} = 8^{x+2}$

Messy Equations

Elizabeth was going to complete her algebra assignment using her graphing calculator. Her older sister Carolyn decided to change the equations that she had copied into her notebook to tease her. Carolyn tells her that the equations are still equivalent to what her teacher had given her. Help Elizabeth by making each equation below quicker and easier to type into her calculator.

1. $y = 2x^{-1}$

2. $y = (2x)^{-1}$

3. $y = \dfrac{7.2x^7}{3.6x^4}$

4. $y = (-4)^0$

5. $y = x^{-4.2}\, x^{6.2}$

6. $y = 3^3 \bullet 3^2$

7. $y = \dfrac{1}{5^{-4}}$

Factoring Quadratics

Sometimes quadratic equations contain hidden information that is revealed when the equations are written in factored form. Change each quadratic equation below to an equivalent equation in factored form.

1. $y = -16t^2 - 48t$

2. $y = 5x^2 - 10x$

3. $y = 3(x - 1) + x(x - 1)$

4. $y = 52x^2 - 13$

5. $y = x^2 + 5x + 4$

6. $y = x(2x + 5) + x(x - 10)$

Digging for Roots

Find any roots or zeros of each quadratic function below. Then explain what roots or zeros mean graphically.

1. $f(x) = 2x^2 - 5x - 3$

2. $f(x) = x^2 + 2x + 1$

3. $f(x) = x^2 + 2x + 3$

4. $f(x) = 2x^2 + 3x - 1$

14

Simplifying Algebraic Fractions

Sometimes it is helpful to express an algebraic fraction in a form that is simpler than the way that it appears. If possible, express each fraction below in a simpler form. If the fraction cannot be expressed in a simpler form, write *not possible*.

1. $\dfrac{x^2 - 4}{x - 2}$

2. $\dfrac{n^2 + 1}{n + 1}$

3. $\dfrac{b^2 + b}{b + 1}$

4. $\dfrac{f + 15}{f^2 + 15f}$

5. $\dfrac{x^2 + 7x + 10}{x + 2}$

6. $\dfrac{n^2 - 8n + 16}{n - 4}$

Olympic Pool Border

An Olympic-size swimming pool measures 25 meters by 50 meters. The Harrison City Recreation Department is planning to construct a new pool that will be bordered with a walkway of a uniform width w. Find the possible widths of the walkway if the total area of the walkway is to be greater than 76 square meters, but no more than 400 square meters.

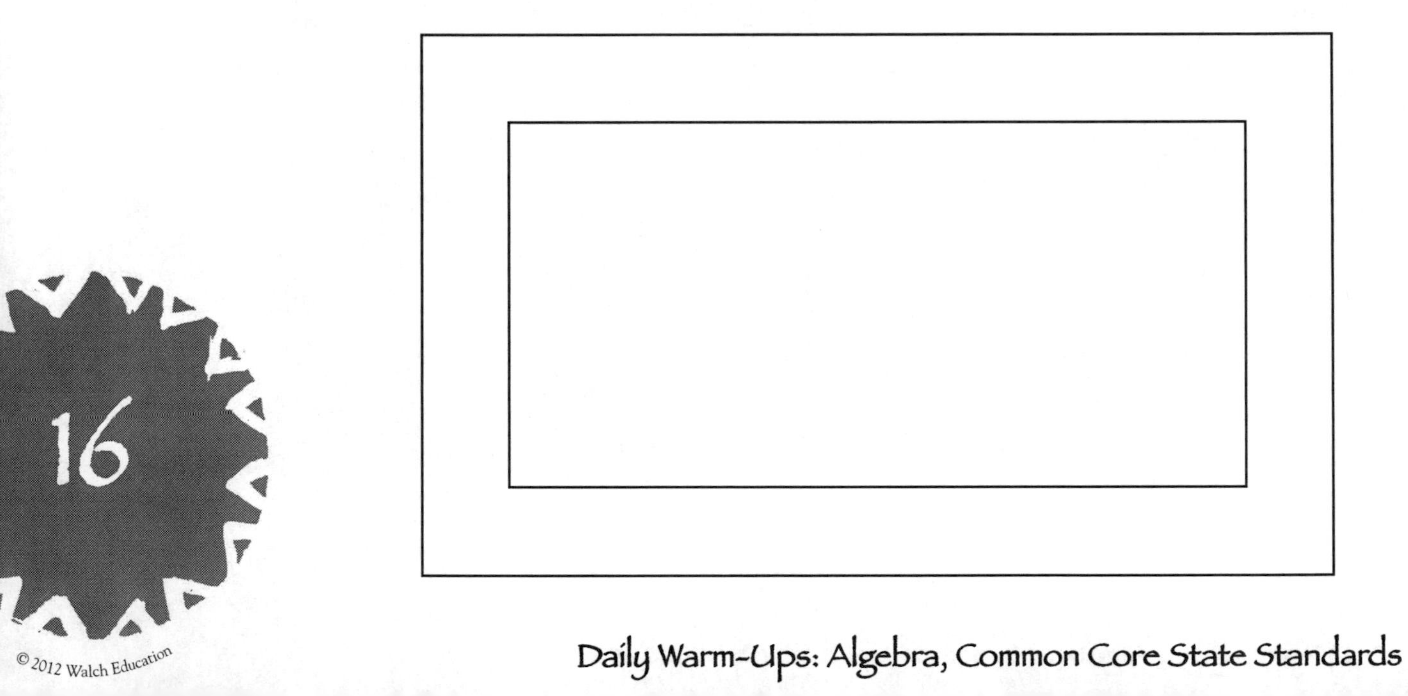

The Seesaw Problem

1. Eric and his little sister Amber enjoy playing on the seesaw at the playground. Amber weighs 65 pounds. Eric and Amber balance perfectly when Amber sits about 4 feet from the center and Eric sits about $2\frac{1}{2}$ feet from the center. About how much does Eric weigh?

2. Their little cousin Aleah joins them and sits with Amber. Can Eric balance the seesaw with both Amber and Aleah on one side, if Aleah weighs about the same as Amber? If so, where should he sit? If not, why not?

Babysitting Bonus

This year, Zachary has been babysitting his young cousins after school for $70 a month. His uncle also gave him an extra bonus of $100 for his excellent work. Since school started, Zachary has earned more than $500. How many months ago did school start? Write an inequality that represents this situation. Solve it showing all your work.

Rolling Along

Did you know that you can calculate the speed (S) of a car by knowing the size of its tires and the revolutions per minute (RPM) that the tires make as they rotate? Think about a car that has tires that are 2.5 meters in circumference. Imagine that the tires are rotating at 500 revolutions per minute.

1. What is the speed of the car?
2. Write verbal and symbolic rules that express the relation between time and distance.
3. Create a table that records some specific data pairs (time, distance) of your choice.
4. What would be the distance traveled in 20 minutes?
5. Sketch a graph that represents the data you have in your table in number 3. Graph the independent variable on the horizontal axis and the dependent variable on the vertical axis.

Graphing Basics

Serena is trying to remember some of the graphs for elementary functions. She is looking at the four graphs shown below. Help her by identifying a sample set of points on each graph. Then find the equation model that should represent each graph.

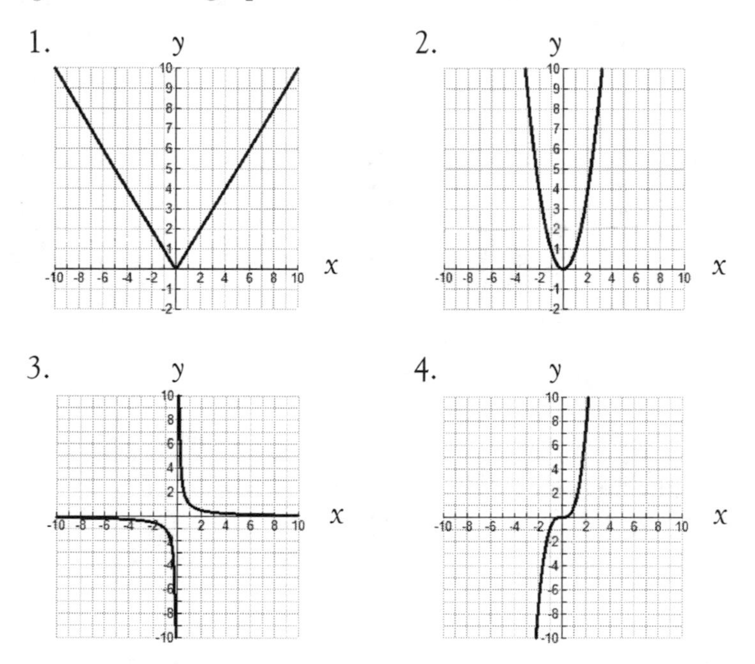

20

Pizza Toppings

You and a group of friends are planning to order two large pizzas and some soft drinks. Each pizza costs \$12.00. Each extra topping costs \$0.50, and each soft drink costs \$1.00. You have a total of \$40.00 to spend.

Represent this situation as a feasible region on a graph, and indicate the corner points of the region and its boundaries.

Solving Equations I

Tucker is tutoring his cousin James in pre-algebra. He is trying to explain different strategies for solving an equation such as $53 = 17 - 4x$. What steps or approaches should he recommend to James? Explain your ideas carefully.

Solving Equations II

Solve each equation below symbolically. Be prepared to justify your answers.

1. $3x + 7 = 19$
2. $7 + 4x = 31$
3. $26 - 3x = 4x + 19$
4. $x^2 - 3.5x = 0$
5. $13x - x^2 = 0$
6. $x^2 - 10x = 5x^2 - 2x$
7. $4x(x - 5) = 0$
8. $3.5(x + 2) + 2(x + 2) = 0$
9. $5.5x = x - 9$
10. $49x - 7x^2 = 0$

Symbol Sense: Original Costs

Think of a real-world situation that might result in each symbolic sentence below. Solve the sentences for C.

1. $C - .25C = \$37.49$

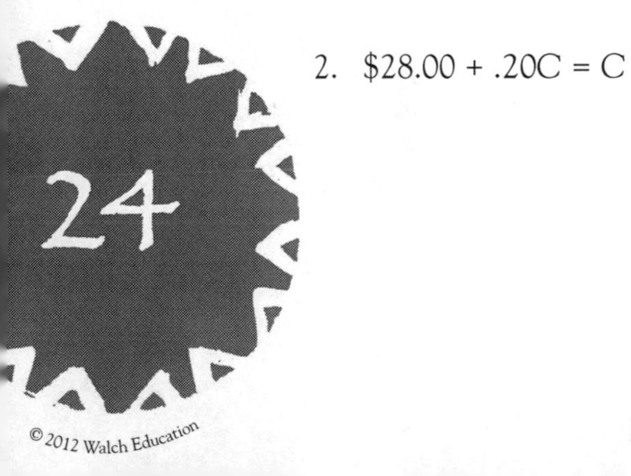

2. $\$28.00 + .20C = C$

Real or Imagined?

Write a description of how you would explain to a friend that $x^2 + 9 = 0$ has no real roots.

Analyzing the Quadratic Formula

Grace has found the following equation in her notebook:

$$x = \frac{-8 \pm \sqrt{8^2 - 4(1)(10)}}{2(1)}$$

She knows that she was using the quadratic formula to solve an equation and is wondering what the original equation might have looked like. Reconstruct her equation. Then find the solution (roots) of the equation. What are the *x*-intercepts for the graph of the equation?

Forms of Quadratic Functions

John and Diana are studying quadratic functions. They have encountered the function below, written in three different forms:

$$f(x) = x^2 - 2x - 3$$

$$f(x) = (x - 3)(x + 1)$$

$$f(x) = (x - 1)^2 - 4$$

Show that the three equations are equivalent. How can you get each form from the other two? What different information is available to John and Diana from each of the equations? What are the zeros, or roots, of the equation? Without graphing and without using a calculator, give the coordinates of the vertex of the parabola and the x- and y-intercepts.

Copy Machine Choices

The Pendleton County School District is trying to decide on a new copier. The purchasing committee has been given quotes on two new machines. One sells for $20,000 and costs $0.02 per copy to operate. The other sells for $17,500, but its operating costs are $0.025 per copy. Which machine would you recommend? Justify your choice with clear mathematics information. How many copies must the school make before the higher price is a reasonable choice?

Nathan's Number Puzzles I

Nathan likes to make up puzzles about integers. Some of his recent puzzles are below. Write symbolic sentences that represent Nathan's puzzles. Then solve each puzzle.

1. Two numbers have a sum of 10. If you add the first number to twice the second number, the result is 8. What are the numbers?

2. One number is twice as large as a second number. The sum of the two numbers is 15. What are the numbers?

3. The first number minus the second number is 2. Twice the first number minus twice the second number is 4. What are the numbers?

Nathan's Number Puzzles II

Nathan has written pairs of linear sentences in symbols. Now he's trying to create word puzzles to go with them. Help him by creating a word puzzle for each pair of linear sentences below. Then find the numbers that fit his equations.

1. $x + 3y = 5$
   $3x + y = 7$

2. $x + 3y = 10$
   $3x + 14 = 2y$

3. $x + 2y = 7$
   $x + 2y = 17$

Systems of Linear Equations I

Marika has asked you to help her understand how to solve systems of equations. Solve each system of equations below using a different strategy. Then explain to Marika why you chose that strategy for that system. Which are best solved by substitution? Which might be easily graphed? Which could be solved by elimination?

1. $y = x - 1$
   $3x - 4y = 8$

2. $3x + 2y = -10$
   $2x + 3y = 0$

3. $x + y = -10$
   $.5x + 1.5y = 5$

4. $3x - 2y = 6$
   $-2x + 3y = 0$

Systems of Linear Equations II

Amal is learning about systems of linear equations. He has come up with some questions regarding equations such as $y = 3x + 8$ and $y = -5x + 11$. Help him by answering the questions below.

1. What is the objective for solving a system of equations such as the one given?

2. How could you find the solution by graphing?

3. How could you find the equation by using tables of values?

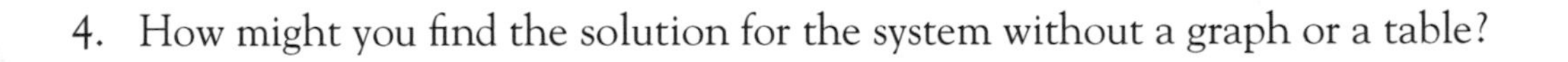

4. How might you find the solution for the system without a graph or a table?

5. What would you look for in the table, the graph, or the equations themselves that would indicate that a system of equations has no solution?

Water Balloon Experiments

Mr. Zeno's algebra class is conducting experiments with water balloons from the roof of the Pendleton County High School. Clarissa is using the formula $h = -4.9t^2 + h_1$, where h is the height of a falling object. She knows that t represents time in seconds and h_1 is the starting height of the object. Her group has found that the height of the high school is 30 meters. Graph Clarissa's information. Using your graph, find out when the balloon will be 10 meters from the ground.

Solving Systems Using Matrices

Theresa has been given the following system of linear equations that models a situation that she is investigating. Create the matrix equation that would also model the situation. Then solve for the two variables.

$$6b + 4s = 1750$$
$$b + s = 400$$

Thinking About Variables, Graphs, and Tables

1. Think about a situation in which variable y depends on variable x. (For example, y might be profit and x the number of items sold.) If y increases as x increases, how would this appear in a table? How would it appear in a graph?

2. Now think about a situation in which variable y decreases as variable x increases. (For example, y might be the amount of gasoline in your car on a trip and x the time you have been traveling.) How would this be indicated in a table? In a graph?

3. In a coordinate graph of two related variables, when do the points lie in a straight line?

4. In a coordinate graph of two related variables, when is it appropriate to connect the points?

The Librarian's Dilemma

Margo is a librarian for the east branch of the Harrison City Library. Each year she adds new books, both fiction and nonfiction, to the library's collection. This year her budget limits her to no more than 75 new books. Library policy states that new fiction books can be no more than half the number of nonfiction books.

1. If x represents the number of nonfiction books and y represents the number of fiction books, write a system of inequalities that models Margo's situation.

2. Graph this system of inequalities.

3. What information is revealed in your graph?

4. What is the greatest number of fiction books that Margo can buy this year?

Graphing Inequalities I

Graph the region indicated by each pair of inequalities below.

1. $y \leq -2x + 7$ and $y \geq 4x + 3$
2. $-3x + 2y > 6$ and $y \leq -2x + 5$

Graphing Inequalities II

Describe the region represented on each graph below using appropriate inequalities.

1.

2.

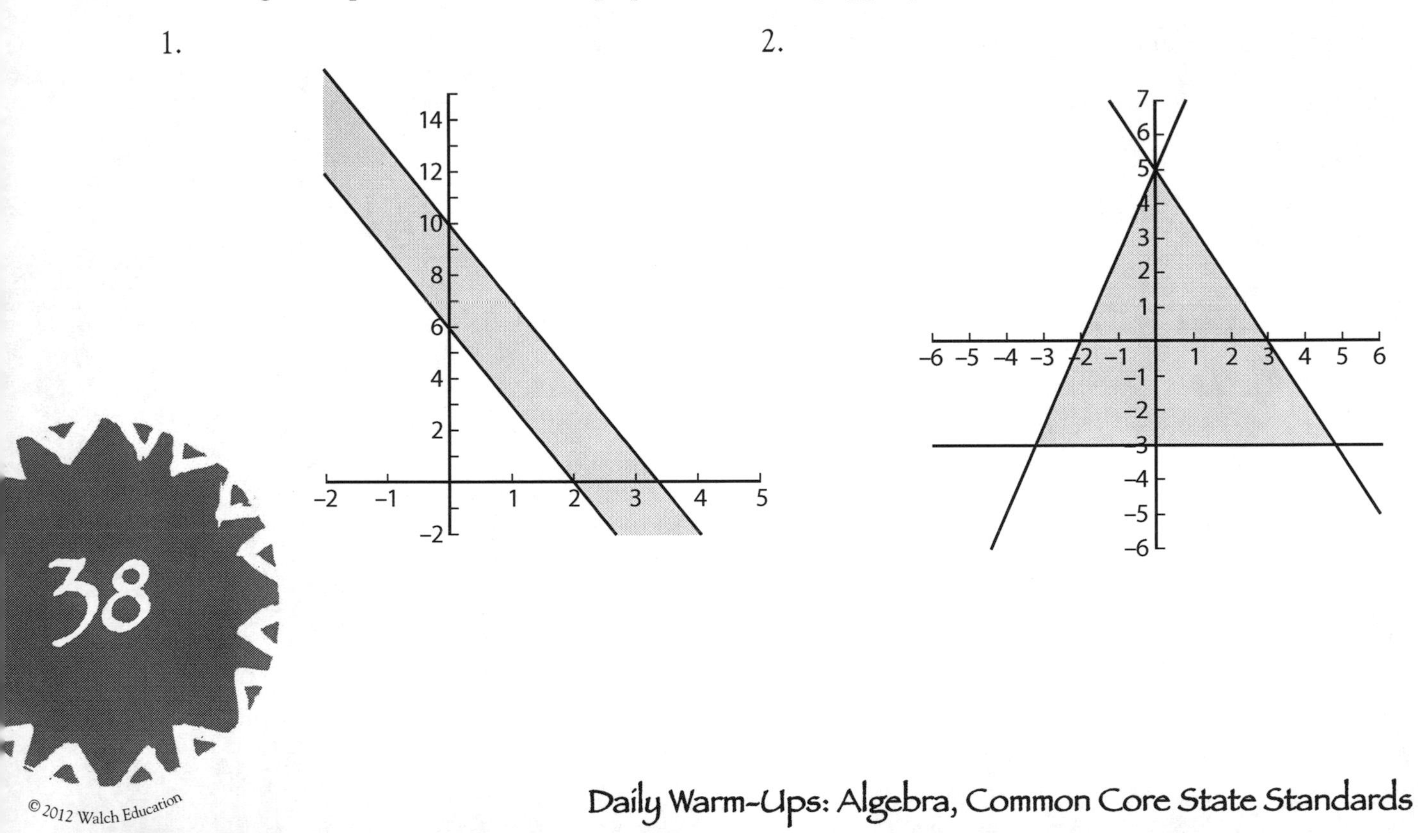

Part 3: Functions

Overview

Interpreting Functions

- Understand the concept of a function and use function notation.
- Interpret functions that arise in applications in terms of the context.
- Analyze functions using different representations.

Building Functions

- Build a function that models a relationship between two quantities.
- Build new functions from existing functions.

Linear, Quadratic, and Exponential Models

- Construct and compare linear, quadratic, and exponential models and solve problems.
- Interpret expressions for functions in terms of the situation they model.

Analyzing Tables

Which tables of (x, y) values given below could represent functional relationships? Explain your thinking.

a.

x	y
0	5
1	7
2	8
3	10
4	5
5	7
6	8
7	9

b.

x	y
3	5
4	9
6	11
3	7
11	9
7	6
8	4
10	6

c.

x	y
4	11
8	7
7	3
9	5
5	12
3	11
2	8
6	9

d.

x	y
a	b
c	d
m	n
a	g
h	s
t	u
r	s
d	k

Cool It!

The temperature in Celsius of a block of hot metal as it cools is given by the function $T(x) = 95(1 - 0.45)^x + 20$, where x is expressed in hours.

1. What is the starting temperature for this hot metal block?

2. What will the temperature be after 3 hours?

3. What will the temperature be when the block has finished cooling?

Oops, Decayed Equation!

During their investigation of radioactive decay, a research group discovered that the equation $f(x) = 200(1 - 0.18)^x$ fits their data. Unfortunately, one member of the group spilled coffee on the table and made the data unreadable. Reconstruct what their table might have looked like using the chart below. Explain what is being represented in the equation by 200 and 0.18.

x	y
0	
1	
2	
3	
4	
5	
6	

Tennis Ball Trauma

A tennis ball has been dropped from the top of a tall building. The ball's height in meters t seconds after it is released can be represented by $h(t) = -4.9t^2 + 150$.

1. Find $h(3)$ and explain what this represents in the situation described.

2. How much time will elapse (to the nearest .01 second) until the ball is at or less than 25 meters above the ground?

3. When will the tennis ball hit the ground?

Fencing Fernando

Franny lives on a small farm with many unusual animals. She has just been given a Vietnamese potbellied pig named Fernando. But before she can get Fernando, she needs to build a pen for him. If needed, she can use one side of a long barn that measures 200 feet. She also knows that she has 120 feet of fencing available. Answer the questions below about Fernando's pen. Be prepared with drawings and mathematical models to justify your answers.

1. What are the longest sides that a rectangular pen could have if the barn is used for one side?
2. Suppose Franny wants a square pen. What are the dimensions of the square, if all the fencing and one side of the barn is used?
3. What is the maximum area that Franny can create for her pig?
4. What is the minimum area that Franny can create?
5. Suppose Franny decides to build her pen in a field, using all of the fencing but not the side of the barn. What are the maximum and minimum areas of her pen?

Maximizing Potato Profit

Maximilian and Minerva have harvested 10,000 pounds of potatoes this fall. They know that they can sell them now for $0.20 per pound. They also know that the price is likely to go up approximately $0.02 per week if they wait to sell. If they wait to sell, however, it is quite possible that they'll lose approximately 200 pounds of potatoes per week due to spoilage and animals munching. When should they sell to maximize their income for their potatoes? Sketch a graph of the relationship and explain your thinking.

Imagining the Graphs of Sequences

Look at the recursive formulas below. Without actually graphing them, imagine the graphs of the sequences generated by these formulas. Then describe each graph using one or more of the following terms: arithmetic, geometric, decreasing, increasing, linear, and nonlinear. Be prepared to justify your thinking.

1. $t_0 = 60$, $t_n = t_{n-1} - 10$, for $n \geq 1$

2. $t_0 = 1{,}000$, $t_n = 0.7t_{n-1} + 100$, for $n \geq 1$

3. $t_0 = 7.5$, $t_n = 0.2 + t_{n-1}$, for $n \geq 1$

4. $t_0 = 98$, $t_n = (1 + 0.25)t_{n-1} - 10$, for $n \geq 1$

5. $t_0 = 75$, $t_n = t_{n-1} \times 1.75$, for $n \geq 1$

Chirping Crickets

Did you know that you can estimate the temperature outside by counting the number of cricket chirps you hear? One method to estimate the temperature in degrees Fahrenheit is to count the number of chirps in 15 seconds and then add 37. This will give you the approximate temperature outside.

1. Write a mathematical sentence that describes the number of cricket chirps per 15 seconds (C) as a function of the temperature (T) in Fahrenheit.

2. Graph this function.

46

3. What do you think are the maximum and minimum values of this function?

Parachuting Down

Cecilia took her first parachute jump lesson last weekend. Her instructor gave her the graph below that shows her change in altitude in meters during a 2-second interval. Use the graph to answer the questions that follow.

1. What is the slope of the line segment?
2. Estimate Cecilia's average rate of change in altitude in meters per second.
3. Give the domain and range for this graph.

Camping Costs

Dan and Julia are planning a camping trip for their school's Outdoor Adventure Club. They went online and found the following information for Hamden Hills State Park. Use the information to answer the questions below.

Number of campsites needed	1	2	3	4	5	6	7
Total campground fee	$17.50	$35.00	$52.50	$70.00	$87.50	$105.00	$122.50

1. Make a coordinate graph of the data. Would it make sense to connect the points on the graph? Why or why not?

2. Using the table, describe the pattern of change in the total fee as the number of campsites needed increases. How is this pattern shown in your graph?

U.S. Population

The graph below shows the United States population from 1900 to 2010, as recorded by the U.S. Census Bureau. Use the graph to answer the following questions.

1. What was the rate of change in population from 1900 to 2000? Is this greater or less than the rate of change in the population from 2000 to 2010?
2. Which 10-year time periods have the highest and the lowest rates of change? How did you find these? What factors might have contributed to these rates of change?
3. What do you predict the U.S. population will be in 2020? Explain your reasoning.

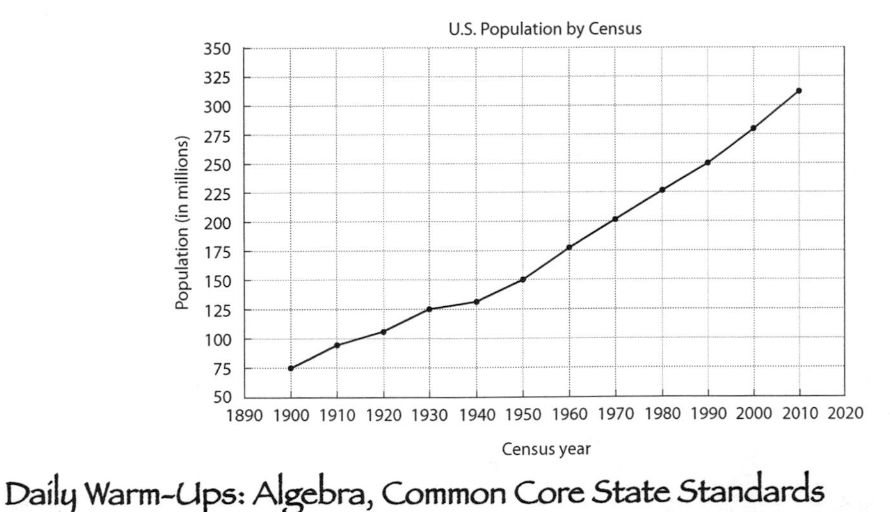

Intervals of Change

Look at the table and graph below.

x	y
–5	–10
–4	0
–2	13
0	16
2	14
4	2
5	–5

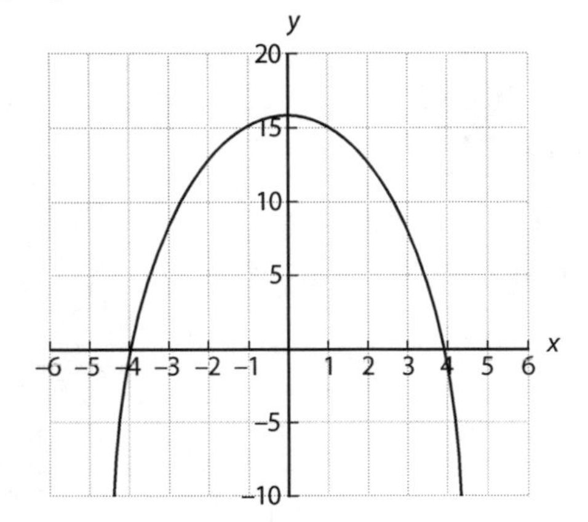

1. Identify the x- and y-intercepts, if any. How is the table helpful? How is the graph useful?
2. Consider the x interval $(-5, 0)$. Describe how y may be changing over that interval of x.
3. Now consider the x intervals $(0, 2)$, $(2, 4)$, and $(4, 5)$ for this graph and table. Is the average rate of change constant over these intervals?

Think About Inverse Power Models

Think about equations of the form $y = \frac{a}{x}$ and $y = \frac{a}{x^2}$. What effect would changes in the value of a have on the graphs of these functions? What patterns would appear in the tables and graphs that could be represented by these equations? Write a few sentences to explain your thinking.

A Ball Is Thrown

Luis threw a baseball straight up at an initial velocity of 120 feet per second. Luis is 6 feet tall. The height of a ball is a function of the time, *t*, that it was thrown and can be found using the following equation:

$$H(t) = -16t^2 + v_1 t + h_1$$

How high did Luis's throw go, and how long did it stay in the air? Create a graph for this situation.

The Profit Is in the Symbols

Molly and Joe are co-chairs of their school's concert committee. They are planning to have a rock band perform at the school. They have been told that the equation $P(t) = -75t^2 + 1500t - 4800$ reflects their potential profit, P, where t represents the price of a ticket. Molly rewrites the equation as $P(t) = (300 - 75t)(t - 16)$. Joe rewrites it as $P(t) = t(-75t + 1500) - 4800$.

1. Are these equations equivalent? What different ideas were Molly and Joe thinking about when they wrote their equations? What might they learn from the equations?

2. Graph the original equation. What are the break-even values for ticket prices for this concert? What is the fixed cost of the concert? According to this equation model, how would the number of tickets sold change if the price was increased? What might be the best price to charge for tickets? Explain your thinking.

Comparing the Graphs

1. Graph the functions $y = x^2$ and $y=\sqrt{x}$ together in the same convenient window. Then compare the graphs. How are they alike? How are they different?

2. What would the graph of $y=\pm\sqrt{x}$ be like? Would this graph still be a function? Why or why not?

Thinking About Quadratic Functions

Quadratic functions are often written in the form $y = ax^2 + bx + c$. Think about how changes in the values a, b, and c affect the graph of the function. Use the graph of $y = x^2$ to demonstrate the changes. Write a few sentences to explain your thinking.

Penny's Parabola

Penny has been given the equation $y = (x - 6)(x + 2)$ to graph. Penny's graph is shown below.

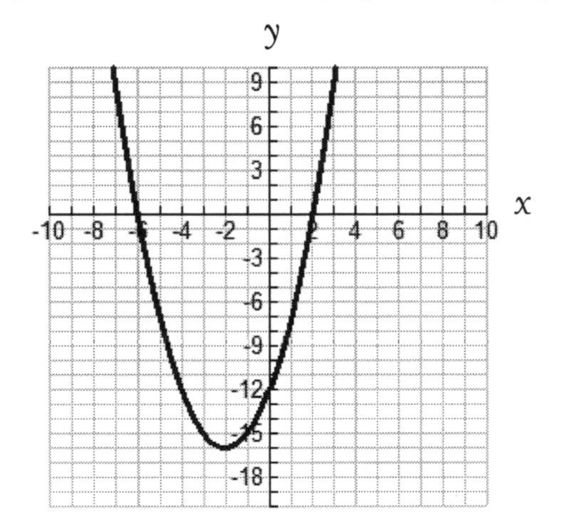

1. Using information about the x-intercepts, the vertex of the parabola, and other forms of the equation of the parabola, tell if you agree or disagree with Penny's graph.
2. Write Penny's equation in polynomial form.

56

A Quadratic Mystery

Without using a calculator to graph these equations, tell which ones are quadratic AND have a minimum point. Explain what you looked for and thought about as you reviewed the equations.

1. $y = x^2 + 5x + 6$
2. $y = x(9 - x)$
3. $y = (x + 3)(x - 2)$
4. $y = -9 - x^2$
5. $y = x(10 + 5)$

A Family of Functions

Consider the following three functions:

$$f(x) = x^2 + 7x + 4$$

$$f(x) = -x^2 + 7x + 4$$

$$f(x) = ax^2 + bx + c$$

What makes them a "family"? In other words, how are they related? What do they look like? Describe their rates of change, symmetry features, maximum or minimum values, and any other special features. What do the values a, b, and c tell you about the nature of the graph in this family?

Discriminating Discriminants

How much do you know about quadratic functions? Without graphing, discuss how many, if any, x-intercepts each quadratic function has.

1. $f(x) = x^2 + 3x + 4$

2. $f(x) = x^2 + 3x - 4$

3. $f(x) = 9x^2 - 6x + 1$

4. $f(x) = 3x^2 + 11x + 6$

Asymptote Adventure

Think about the functions given below. Determine the domain, range, and all asymptotes for each function's graph.

1. $f(x) = 7 + \frac{3}{x}$

2. $g(x) = \frac{1}{x-5} + 3$

3. $h(x) = \frac{2x + 3}{x}$

It's All in the (Rational) Family

Consider the following functions:

a. $r(x) = \dfrac{3}{x}$

b. $r(x) = \dfrac{3}{x^2}$

c. $r(x) = \dfrac{3}{x^3}$

d. $r(x) = \dfrac{3}{x^4}$

1. Predict what the graphs of these functions will look like without graphing them first. How will they be alike? How will they be different? What are their respective asymptotes? Is it appropriate to call them *functions*? Why or why not?
2. Now graph the functions using graph paper or a graphing calculator. Do you notice any additional characteristics?

Exponential Function Features

1. Consider the following functions:

   a. $g(x) = 3 \bullet 2^x - 1$

   b. $h(x) = 2 \bullet 3^x + 4$

   c. $k(x) = 2 \bullet \left(\frac{3}{2}\right)^x$

Without graphing the functions, make some predictions based on your understanding of exponents. What happens to the graph of each function if x gets very large? What if x is negative? Where do these graphs cross the y-axis?

62

2. Now consider the function $f(x) = a \bullet b^x + c$. How do the values of a, b, and c affect the graph of the function? What if $a < 0$? What if $0 < b < 1$? What happens when you change b to $(\frac{1}{b})$?

Functioning Properly?

Gwen and Rita are discussing the characteristics of certain functions. Rita claims that the equation $y = x^2$ does not model a function because the line $y = 5$ intersects the graph of the equation in two points. Gwen disagrees. How might she present her case to Rita?

Another Functional Family

Consider the following three functions:

$h(x) = 3^x$

$h(x) = (0.3)^x$

$h(x) = a^x$

What makes them a "family"? In other words, how are they related? What do they look like? Describe their rates of change, symmetry features, maximum or minimum values, and any other special features. What features make this family different from other families? What effect do the constants have on the function and its graph?

The Step Function

Draw a graph of $y = [x + 2] + 1$. Why do you think this function is called a "step" function? What other name is applied to this type of function? Why?

Thinking About Holes and Asymptotes

Some equations create unusual results when they are graphed. Think about equations and graphs that might have a hole or an asymptote at $x = 2$. Give some examples. Then write a few sentences explaining how you can use an equation to predict where there would be a hole or an asymptote. How can you tell how to write an equation for a function from a graph with a hole or an asymptote?

Windows of Opportunity

Ben, Carla, and Ted are exploring the graph of $f(x) = \dfrac{2x^2 + 13x + 7}{x - 2}$. Ben believes that it represents a parabola. Carla thinks that it looks like a linear function. Ted doesn't know what to think. They have asked you to help them. The graphs they have drawn are shown below.

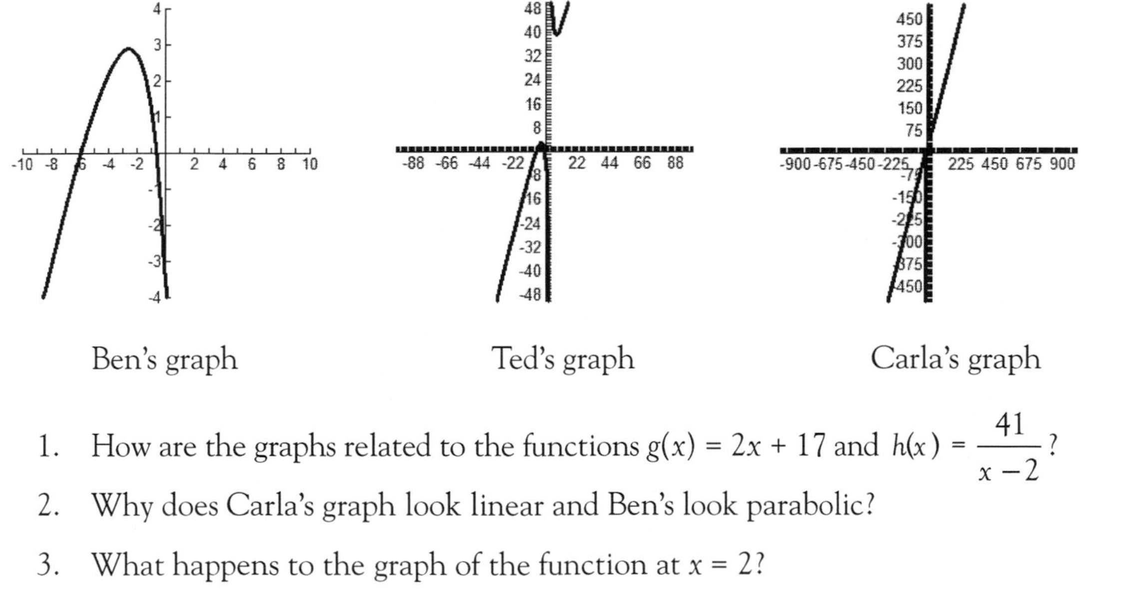

Ben's graph Ted's graph Carla's graph

1. How are the graphs related to the functions $g(x) = 2x + 17$ and $h(x) = \dfrac{41}{x - 2}$?
2. Why does Carla's graph look linear and Ben's look parabolic?
3. What happens to the graph of the function at $x = 2$?

Exploring Constant Areas

Juan is exploring the relationship between the widths and lengths of rectangles with a constant area. He states that this relationship should be represented as an inverse function. He has been considering rectangles with an area of 24 in^2 and rectangles with an area of 36 in^2. Do you agree or disagree with Juan's statement? What tables of data might he have? How might he be representing the situation symbolically and graphically?

Granola Inflation

A box of Crunchy Granola is selling for $2.98 at the Healthy Grocery Mart. The inflation rate for granola has been noted to be 4% annually.

1. What do you expect the price to be a year from now?

2. What would you expect the price to have been a year ago?

3. Write an equation that models this information. Does this represent exponential growth or decay?

4. Assuming a steady inflation rate, in what year would you expect the price of a box of Crunchy Granola to be more than $3.50?

Maximum Area

Sam and Jenny have 1,200 feet of fence. They would like to enclose as large a rectangular plot of land as possible. Their farm lies along a river, and they will not need to put up a fence along that side of their land. What is the maximum area that they can get from the length of fence that they have?

Mystery Table

Sophia and Tran have discovered this partially completed table in their algebra notes. Sophia thinks it represents a quadratic function. Tran thinks it represents exponential decay. Do you agree with either of them? If so, who and why? If not, why not? Does the table represent some other kind of function? Complete the table and write a symbolic sentence for the function that you think is represented by the table values.

x	y
–2	
–1	
0	64
1	12.8
2	2.56
3	.512
4	

DVD Rentals

Rosa's parents just bought a new DVD player for the family. Rosa's mom asked her to research rental prices for DVDs online. More Movies has a yearly membership package. Rosa found the following table of prices on the company's web site.

More Movies DVD Rental-Membership Packages

Number of videos rented	0	5	10	15	20	25	30
Total cost	$30	$35	$40	$45	$50	$55	$60

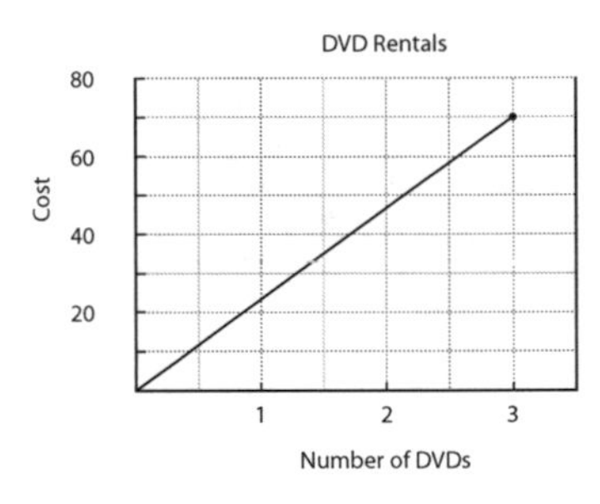

On the Deluxe DVD Rentals web site, Rosa found that there was no membership package offered. She made the graph on the right to show how the cost at Deluxe DVD Rentals is related to the number of videos rented.

72

If both rental services have a comparable selection of DVDs and Rosa's family will watch on average about two movies a month, which service should they choose? Explain how Rosa's family might decide which service to use. Using the information about each service, describe the pattern of change relating the number of DVDs rented to the total cost.

Boxes, Little Boxes

Jaden is exploring the characteristics of small boxes. He is creating boxes using one sheet of construction paper for each box. He makes the box by cutting out squares in 1-inch increments from the four corners of the paper and then folding up the edges. If the paper that Jaden is using is 9 inches by 12 inches, what are the sizes of the boxes that he is designing? Create a table of values that records the length, width, and volume of all the boxes you think Jaden could create in this way. Also include the length of the square cutouts, the surface area of the box, and any other data of interest to you. Do you see patterns in your data? What is the length of the square cutouts for the box with the greatest volume? With the smallest volume? If s represents the side of the square that Jaden cuts, how could he represent the volume of the resulting box in terms of s?

The Growing Cube

A cube is having its first birthday. This cube is called a *unit cube*. Describe how many corners (vertices), faces, and edges it has. On the cube's second birthday, its edges are twice as long as on its first birthday. How many unit cubes would be needed to equal the volume of this new 2-year-old cube? On the cube's third birthday, its edges are 3 times as long as those of the unit cube. This pattern continues for each birthday. How many unit cubes would be needed to equal the volume of a cube that is 3, 4, 5, 6, 7, 8, 9, and 10 years old? Write a rule to calculate the number of unit cubes needed for any number of birthdays. Then draw a graph of the number of unit cubes as a function of the number of years old the cube is.

Switching Chips

Rocco is exploring a children's puzzle that involves a strip of 7 squares, 3 white chips, and 3 black chips.

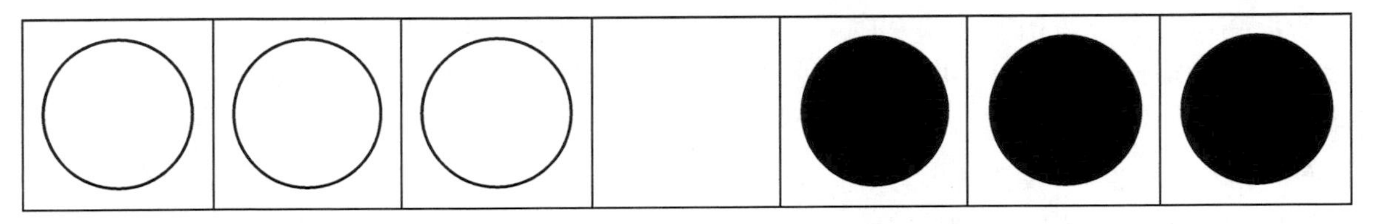

The object of the puzzle is to switch positions of the chips so that all the black chips are on the left and all the white chips are on the right. A chip can move by sliding to an adjacent empty square or by jumping over one chip to land on an empty square. White chips only move right, and black chips only move left.

1. How many moves will it take to switch the chips?
2. Complete the table to show the number of moves it would take for other numbers of chips.

Number of each color	1	2	3	4	5	6	7	8	9	10
Number of moves										

3. A puzzle with n white chips and n black chips will require how many moves to complete the switch?

Fair Trade

Land developers sometimes make deals with landowners and trade pieces of land that they need for other plots of land in other places. Suppose you own a square piece of property that the ABC Land Development Corporation would like to have to build a new mall. The company is willing to trade your property for a rectangular piece nearby that is 3 meters longer on one side and 3 meters shorter on the other side. How does this rectangular piece compare to your original square plot? Would this be a fair trade for any side length of your original square? Complete the table and compare the areas. Write functions for the situation.

Original square		Rectangular plot			Difference in area
Side length	Area	Length	Width	Area	
4	16	7	1	7	9
5					
6					
7					
8					

The Shape-Shifting Square

A square has a side of length s. A new rectangle is created by increasing one dimension by 5 centimeters and by decreasing the other dimension by 4 centimeters.

1. Draw a sketch that represents this situation.

2. Write functions that represent the areas of the two figures.

3. For what values of s will the area of the new rectangle be greater than the original square?

4. For what values of s will the area of the new rectangle be less than the original square?

5. For what values of s will the areas of the two figures be equal?

6. Explain how you found your answers.

Ping-Pong Prices

Lily found the price of table tennis balls listed on the Internet at $4.75 for a package of 6 balls. Shipping and handling was listed at $1.00 per package.

1. Write an equation that represents the total cost for different numbers of packages of table tennis balls.

2. Sketch a graph of this relationship.

3. If you shift your graph up a value of $0.50, does this mean the price per package increased, or the shipping price increased?

4. Write a new equation for the situation in question 3.

Hedwig's Hexagons

Hedwig is using toothpicks to build hexagon patterns. The first three are pictured. How do you think her pattern will continue? She has begun to collect information in a table in order to explore the relationships among the number of toothpicks, the number of hexagons, and the length of the outside perimeter of the figure. Complete the table for her.

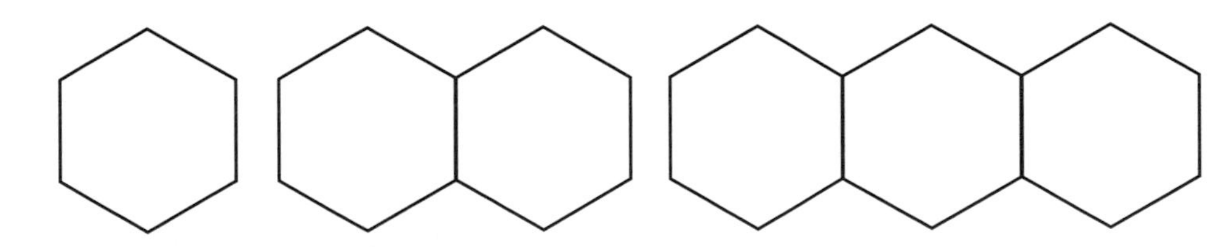

Number of toothpicks			
Number of hexagons	1	2	
Perimeter of figure		10	

Now write the function for the relationship between the number of hexagons in each figure and the perimeter of the figure. Write another function for the relationship between the number of hexagons and the number of toothpicks.

Triangular Numbers

Triangular numbers can be represented using dots, as shown below. The first triangular number is 1, the second is 3, the third is 6, and the fourth is 10, as pictured. How many dots will be in the fifth and sixth figures? Can you predict how many dots will be in the *n*th figure? Write a function that could be used to determine the *n*th figure.

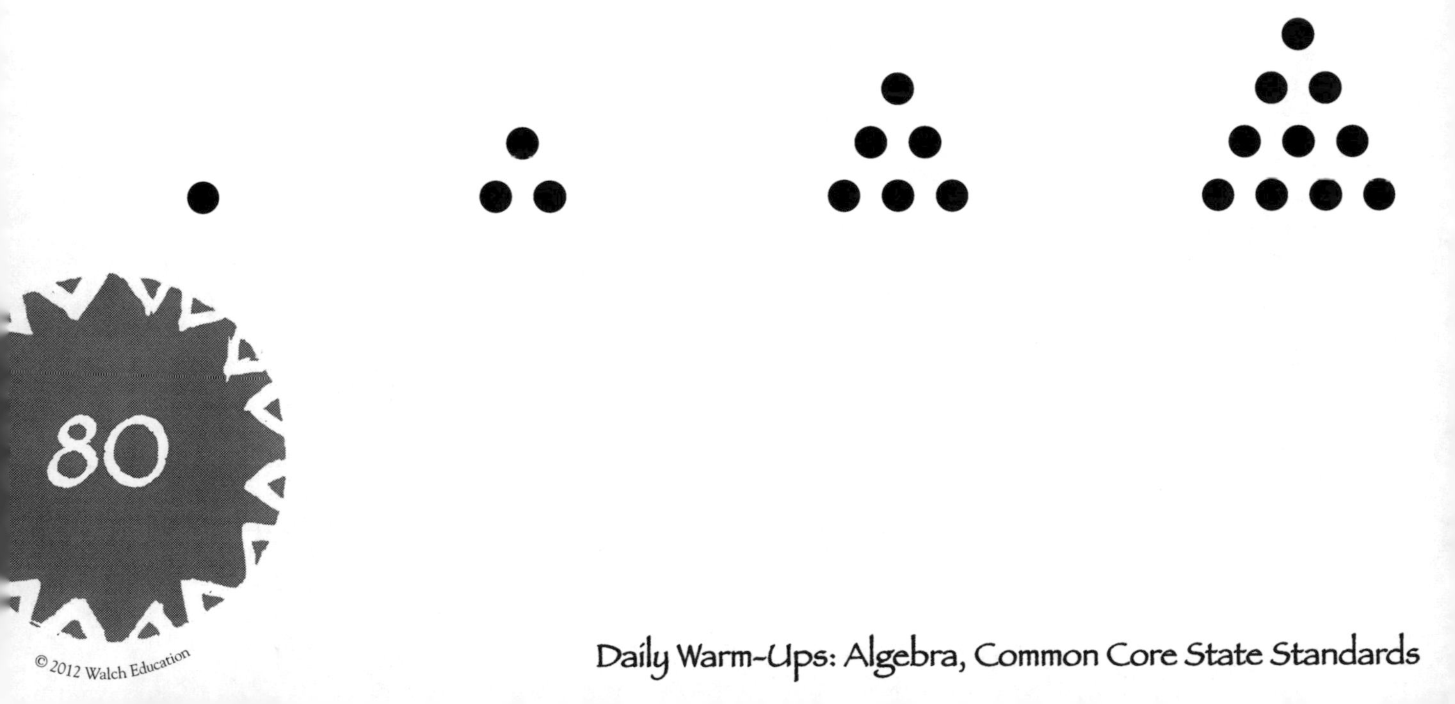

Patterns with Squares

Alexandra is baby-sitting her little brother. To entertain him, she is building structures with his blocks. The picture below represents three of the structures that she has built.

1. Sketch pictures of what you think the next two structures will look like if Alexandra continues the pattern that she has started. What number pattern is suggested by the number of blocks used in this sequence of structures?

2. How many blocks would be in the tenth structure if the pattern continues?

3. How many blocks would be in the 100th structure?

4. How many blocks would be in the *n*th structure?

Pentagonal Numbers

Pentagonal numbers can be represented using circles to form a pentagonal array as in the diagram below. Find a generalization for the *n*th pentagonal number.

P_1 P_2 P_3 P_4

Newspaper Numbers

Many newspapers consist of large sheets of paper that are folded in half to form the pages. For example, a newspaper that is made from 2 large sheets of paper would have 8 pages. Consider the following questions about newspapers created in this way.

1. If there were 10 sheets of paper, how many pages would there be?
2. What page numbers would appear in the center, or on the innermost sheet of the newspaper? What is the sum of the numbers?
3. What is the sum of the page numbers that would appear on ANY one side of ANY sheet of the newspaper?
4. What is the sum of all the pages of the newspaper?
5. Write a rule in words or symbols that would give the sum of the pages for a newspaper with N sheets of paper.

Stacking Cereal Boxes

Freya works part-time in her family's grocery store. Her dad has asked her to stack 45 cereal boxes in a display area for an upcoming sale. The boxes have to be stacked in a triangle with 1 fewer box in each row ending with 1 box at the top (like the sample stack shown below). How many boxes should Freya put in the bottom row? Suppose she had *B* boxes to stack in the display. Help her find a rule or formula to determine how many boxes she should put in the bottom row.

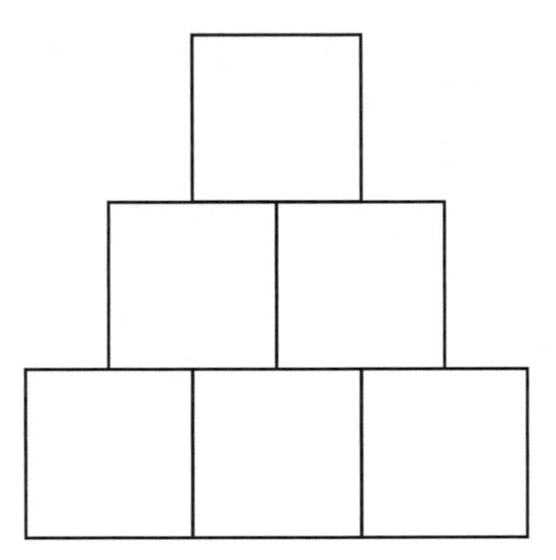

84

Pizza Cutting

Paolo loves pizza. He even dreams of pizza and makes up math problems about pizza. He wonders how many pieces of pizza he could get by cutting a very large circular pizza exactly 7 times. He decides that it doesn't matter if the pieces are the same size. He also decides that it isn't necessary for the cuts to go through the center of the pizza. How many pieces of pizza could he make? What if Paolo made n cuts in the pizza?

What is the greatest number of pieces he could make? Use the table below to help you.

Number of cuts	0	1	2	3	4	5	6	7	8	9			n
Number of pieces	1	2	4	7									

85

An Odd Pyramid

Look for patterns in the pyramid below. Then answer the following questions.

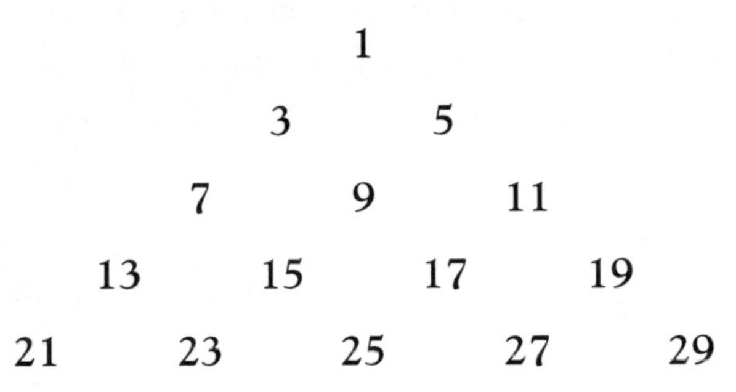

86

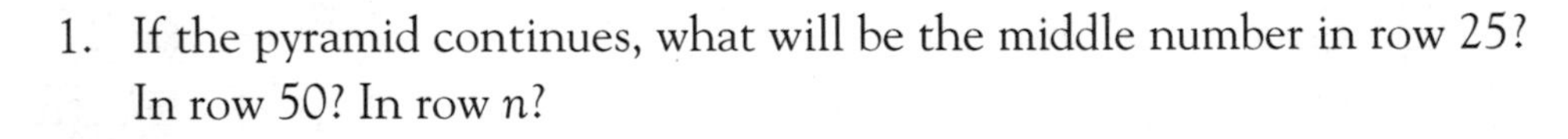

1. If the pyramid continues, what will be the middle number in row 25? In row 50? In row n?
2. What is the difference between the first and last number in row 4? In row 5? In row 40? In row n?
3. What is the last number in row 5? In row 6? In row n?
4. What is the sum of the numbers in row n?
5. What is the sum of all the numbers *through* row n?

How Long? How Far?

The Carmona family lives in Minnesota. They are driving to Florida for a vacation at an average speed of 60 miles per hour. Write an equation for a rule that can be used to calculate the distance they have traveled after any given number of hours. Then write a brief letter to the Carmona family that describes the advantages of having an equation, a table, and a graph to represent their situation.

Pool Border

Landscapers often use square tiles as borders for garden plots and pools. The drawing represents a square pool for goldfish surrounded by 1-foot square tiles.

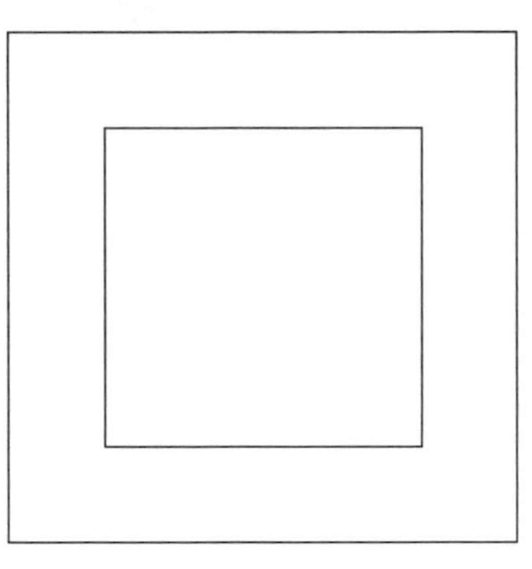

1. How many tiles will be needed for the border of this pool with an edge of length S feet?
2. Express the total tiles in as many different ways as you can. Be ready to explain why your different ways are equivalent.

Game Time

Abby and her brother Harry like to play a game called "U-Say, I-Say." Harry gives the "U-Say" number (an integer between −10 and +10). Abby has a secret rule she performs on the number that results in the "I-Say" number. Complete the table below by giving the missing "I-Say" values. Then describe Abby's rule in words and symbols.

U-Say	3	0	–4	1	2	5
I-Say	11	2	–10	5		

Coloring Cubes

Crystal and T.J. are considering cubes of many sizes that have been dipped completely in paint. They start by thinking about a cube that is 10 units long on each side. After it is dipped and it dries, the 10-unit cube is cut into single-unit cubes.

1. How many of these newly cut unit cubes are painted on 3 faces, 2 faces, 1 face, and zero faces? Answer those same questions for cubes that are 2, 3, 4, 5, 6, 7, 8, and 9 units long on each side. Then make a chart of your findings for all the cubes. Look for patterns.

2. Where are the unit cubes with 3 painted faces located on the various cubes? With 2 painted faces? With 1 painted face? With zero painted faces?

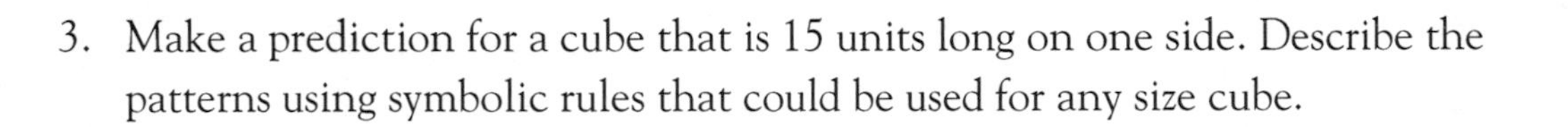

3. Make a prediction for a cube that is 15 units long on one side. Describe the patterns using symbolic rules that could be used for any size cube.

Credit-Card Charges

Charles purchased a new flat-screen television for $2,000 with his credit card on July 1. His credit-card company charges 1.5% interest on the unpaid balance.

1. If Charles does not pay the credit-card company at the end of one month, how much will he owe?

2. How much will he owe at the end of two months?

3. How much will he owe at the end of three months?

4. When will his unpaid balance reach $2,500?

Depreciating Car Values

Lupita wants to buy a car that will have the best resale value after 3 years. At one dealership, she has found a car she likes that sells for $15,000 and depreciates at a rate of 30% per year. At another dealership, she has found a car that costs $12,000 and depreciates at a rate of 20% per year. Which car will have the greater value in 3 years?

Bouncing Ball

A hard rubber ball will rebound to 75% of its height each time it bounces. If the ball is dropped from a height of 200 centimeters, what will the height of each bounce be until the ball rests? Create a table and a graph of the ball's bounce rebound height over several bounces. On which bounce will the rebound be less than 50 centimeters?

Looking for Lines I

Linear patterns can be recognized in several different representations of a relationship between two sets of quantities. Sometimes we look for the relationship in graphs, tables, or symbolic rules. Describe how you can tell whether a situation is or can be a linear model by looking at the information below.

1. a scatter plot of the data

2. a table of the values

3. the equation of the relationship

4. a description of the problem

Identifying Graphs from Equations

Equations can reveal a lot of information. You can tell a lot just by how an equation looks and the variables and operations that are in it. Based on what you know about the nature of equations, identify whether each equation below represents a quadratic, a linear, or an exponential relationship. Explain how you made each decision.

1. $y = 7x + x^2$

2. $y = 3x + 7$

3. $y = (6 - x)x$

4. $y = 3^x$

5. $y = 3x(x + 4)$

Spring Break Flight Costs

Ray Airhart is chief pilot for Eagle Valley Aviation. He is promoting a charter flight to Florida for the Eagle Valley High School senior class spring break. The flight will cost $200 per person for 10 students or fewer. For each additional person who goes on the trip, he can lower the price for each person an additional $4. Construct a function that models the cost per person over the domain, $x \geq 10$.

Phone Plan Decisions

Aisha's parents are planning to give her a new cell phone for her upcoming birthday and pay for her service for one year. They have asked her to determine which plan is best for her. Best Talk offers service at a monthly basic fee of $20.00 per month and $0.10 per minute for each minute used. Horizon One-Rate has no monthly fee, but charges $0.40 per minute. Create a table, a graph, and equations that model this situation. Then give an argument for why Aisha might choose either plan.

Used Car Purchase

Mary Ellen has just started a new job and will need a car to drive to work. She stopped by Jolly Joe's dealership and found a car that she really likes. Joe has told her that she can buy the car for a down payment of $2,000 and a monthly payment of $399 for 24 months.

Happy Hank runs a competing dealership nearby. He has a car almost exactly like the one Mary Ellen wants. Hank will sell her the car for a down payment of $1,600 and a monthly payment of $420 for 24 months.

1. Write an equation for each situation that represents the amount that Mary Ellen will have paid up to that point for any month in the 24-month period.
2. What amount will she have paid at the end of one year if she buys Jolly Joe's car? What if she buys Happy Hank's car? What will she have paid for each car at the end of two years?
3. Which is the better deal?

98

Slope-Intercept Equations

Find a function written in slope-intercept form that satisfies each condition below.

1. a line whose slope is -3 and passes through the point $(2, 5)$

2. a line whose slope is $\frac{-2}{3}$ and passes through the point $(4, -1)$

3. a line that passes through the points $(2, 6)$ and $(6, 1)$

4. a line that passes through the points $(-2, 3)$ and $(4, -3)$

5. a line that passes through the points $(-1, -1)$ and $(-5, -5)$

Up and Down the Line I

Write a function for the line pictured in each graph below.

1. 2. 3. 4.

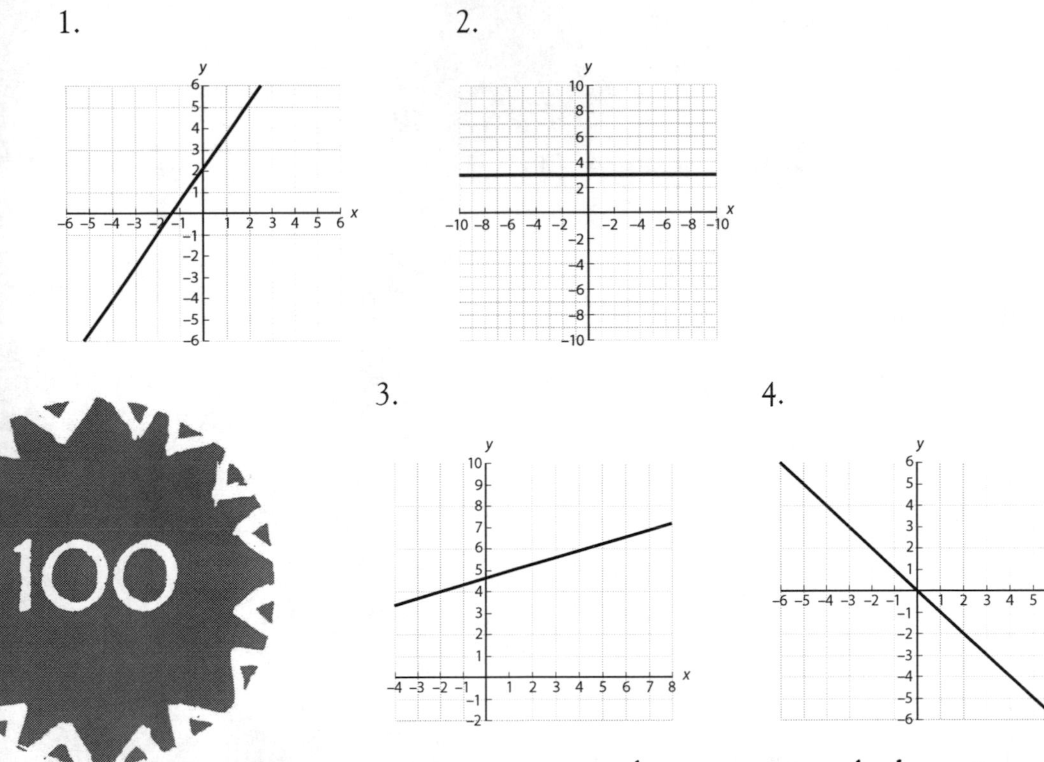

Up and Down the Line II

Write a linear function for each condition below.

1.

x	−1	0	1	2	3
y	1	3	5	7	9

2. a line whose slope is –3 and y-intercept is 5

3. a line that passes through the points (2, 5) and (5, 6)

4. a line that passes through the point (3, 7) and has a slope of $\frac{2}{3}$

Butler Bake Sale

The Butler High School sophomore class is planning a bake sale as a fall fund-raiser. Luis is chairing the planning committee. He gave a brief survey to determine what price should be charged for each brownie. He predicts from his results that a price of $0.50 per brownie will result in 200 brownies sold. He also predicts that a price of $1.00 per brownie will result in about 50 brownies sold. He is assuming that the relationship between the brownie price and the number sold is a linear relationship.

1. Write a function for the relationship that Luis has predicted between the cost and the number of brownies sold.

2. Find the slope and *y*-intercept. Then explain what these mean in the context of Luis's information.

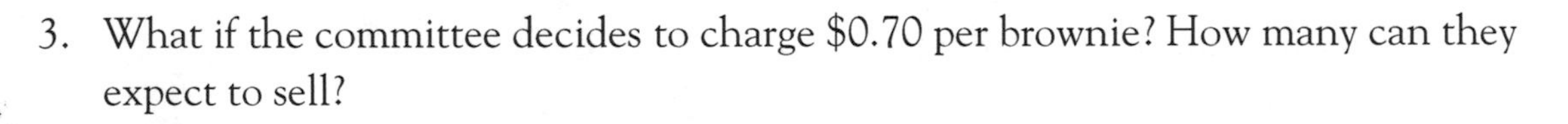

3. What if the committee decides to charge $0.70 per brownie? How many can they expect to sell?

4. Luis has taken an inventory and found that the class has 300 brownies to sell. Using your equation, what would be an appropriate price to charge for each brownie?

102

Looking for Lines II

Explain how you can find the equation of a line if you know the information below. Use examples to explain your thinking.

1. the slope and y-intercept

2. two points on the line

3. the slope of the line and a point that is on the line, but is not the y-intercept

Thinking About Equations of Linear Models

Think about ways to find an equation of the form $y = mx + b$ or $y = a + bx$ from a table of data or a graph of the points. How can you find the equation if you know the slope and y-intercept? How can you find the graph by looking for the rate of change and other values from the table? How can you find the equation of the line if the slope and y-intercept are not given? Write a few sentences to explain your thinking.

Freezing or Boiling?

Lucas knows that the relationship between Celsius and Fahrenheit temperatures is linear. However, he often forgets the equation. He remembers that at freezing, the temperatures are (0°C, 32°F). At boiling, the respective temperatures are (100°C, 212°F). Explain to Lucas how he can find this equation knowing that the relationship is a linear one.

Weekly Pay Representations

Amanda has a part-time job working for a local company selling wireless phone subscriptions. She is paid $120 per week plus $25 for each subscription that she sells.

1. Complete the table below, showing Amanda's weekly pay as a function of the number of subscriptions that she sells.

Number of subscriptions sold	Weekly pay (dollars)
1	
2	
3	
6	
10	
12	
15	

2. Using variables, write two rules (equations), one for the weekly pay in terms of her sales, and one for her sales in terms of her pay.

3. Amanda wants to earn at least $600 next week. How many subscriptions must she sell?

Swimming Pool Bacteria

1. At the local swimming pool, the bacterial count is measured on Monday morning at 8 A.M. and found to be 1,200 bacteria per cubic centimeter. If the count is thought to double every 24 hours, what will the bacteria count be on Thursday afternoon at 2 P.M.?

2. If the safe range of bacteria is less than 200,000 per cubic centimeter, when will this pool need to be treated? Give an equation and a graph with your solution.

Thinking About Change and Intercepts in Linear Models

Think about linear models in various forms. How can you see or find the rate of change in a linear model from a table of values? How would you find it from the graph? How does it appear in the equation? How would you determine the y-intercept in those three situations? Write a few sentences to explain your thinking.

Modeling Tidal Cycles

Melinda has created a trigonometric equation. She believes the equation models the height of water on the pylons of her father's dock during the tide cycle. Her equation is $h = 7.5\cos\left(\frac{2\pi(t-3)}{12}\right) + 16$, where t represents the time in hours since midnight and h represents the height in feet. Melinda's father wants to launch his boat when the water is approximately 12 feet high on the pylons. He thinks that this will happen around 8:00 A.M. What do you think about this prediction? Using Melinda's equation model, predict the first time when the height of the water will be approximately 12 feet on the dock pylons. When will the water be that height again?

Part 4: Statistics and Probability

Overview

Interpreting Categorical and Quantitative Data

- Summarize, represent, and interpret data on two categorical and quantitative variables.

Water Balloon Bungee Jump

Natalie and Charlotte are conducting an experiment using rubber bands and water balloons. Their challenge is to predict the distance a water balloon will fall for any number of rubber bands that they might use. This is the data that they have collected so far:

Number of rubber bands	3	4	5	6	7	8	9	10	11
Distance fallen (cm)	71	83	95	115	128	144	160	175	190

They plan to test their conclusions from this data by dropping a water balloon from a stairwell at a starting height of 12.5 feet. How many rubber bands should they use to get the water balloon as close to the floor as possible, but not break it?

Pet Mice

Linn has been keeping several mice as pets for some time. She has recently learned that she can raise mice to sell to pet stores. However, she would need to provide the mice in lots of at least 200. To predict when she will have enough mice to sell, she has collected the data in the chart below.

Month	Mice
1	3
2	7
3	17
4	36
5	65

1. Make a scatter plot of Linn's data and draw a line of best fit.

2. Find an equation in slope-intercept form for your line.

3. Is Linn correct in using a linear model to predict when she will have 200 mice? If so, predict when. If not, explain why not.

Relating Temperature and Altitude

Keri has just begun pilot lessons. She is very interested in the relationships among altitude, pressure, temperature, and density. She did some online research and found the table below. It gives the standard values of pressure, temperature, and density at different altitudes.

Altitude (ft)	Pressure (Hg)	Temperature (°F)	Density (%)
sea level	29.92	59.0	100
2,000	27.82	51.9	94.3
4,000	25.84	44.7	88.8
6,000	23.98	37.6	83.6
8,000	22.22	30.5	78.6
10,000	20.57	23.3	73.8
12,000	19.02	16.2	69.3
14,000	17.57	9.1	65.0
16,000	16.21	1.9	60.9

112

Use a graphing calculator to make a graph of the relationship between altitude and temperature for $0 \leq A \leq 20{,}000$. Find a regression equation that you think works for this data. Be ready to explain your thinking.

Life Expectancy

The table below shows the average number of years a person born in the United States can expect to live based on their year of birth. What do you predict is the life expectancy of a man, a woman, or any person born in 2020? Find a line of best fit for this data. Write an equation for the line you choose. Make your prediction using your graph and your equation.

Year	Female	Male	Both sexes
1930	61.6	58.1	59.7
1940	65.2	60.8	62.9
1950	71.1	65.6	68.2
1960	73.1	66.6	69.7
1970	74.7	67.1	70.8
1975	76.6	68.8	72.6
1980	77.4	70.0	73.7
1985	78.2	71.1	74.7
1990	78.8	71.8	75.4
1995	78.9	72.5	75.8
2000	79.7	74.3	77.0
2005	80.4	75.2	77.9

Grow Baby!

The average growth weight in pounds of a baby born in the United States is a function of the baby's age in months. Look at the sample data provided and graph the information in an appropriate window. Then write two sentences about the graph and the relationship between the two variables.

Age in months	Weight in pounds
0	7
3	13
6	17
9	20
12	22
15	24
18	25
21	26
24	27

Braking/Stopping Distances

Use a graphing calculator or graph paper to graph the relationship between miles per hour and stopping distance using the data from the table below. What questions come to mind as you look at your graph? Make a prediction for the stopping distance at 80 miles per hour. Explain and justify your prediction.

Miles per hour	Stopping distance
10	27
15	44
20	63
25	85
30	109
35	136
40	164
45	196
50	229
55	265
60	304
65	345

Thinking About Linear Models

Imagine that two of your classmates have proposed two different linear models for a table of data that you are investigating. This means that they have given you two different lines and two different equations for the situation. How can you compare the two models and the data to find out which is the best model for the situation? Write a few sentences to explain your thinking.

Mr. Wiley's Baby

Mr. Wiley presented some data to his algebra class about the early growth of his new baby boy. Hermione created a scatter plot for Mr. Wiley's data. The data and scatter plot are shown at right.

Week	Weight
1	8.5
2	9.25
3	9.75
4	9.75
5	10.5
6	11
7	11.5
8	11.75

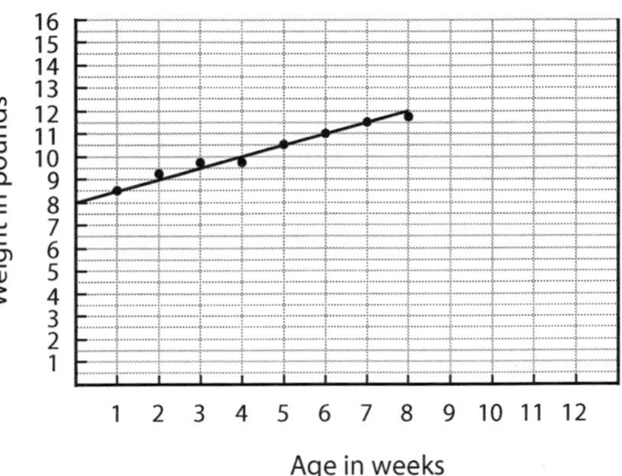

1. Does the data seem to represent a linear model? What equation fits Hermione's graph model?

2. How long do you suppose this growth pattern will continue in this way? If it does continue, what would you expect Mr. Wiley's son to weigh in 1 year? In 2 years? In 14 years?

3. Within what age limits do you think this is a reasonable representation of the growth of Mr. Wiley's baby?

Explain Linear Regression

Write a brief but clear description of the conditions that would prompt you to pursue a linear regression equation on your graphing calculator. Explain your thinking. Describe what you might see on a graph of data points. Tell what keystrokes would be necessary to find the linear regression.

Table Patterns I

Look at the tables below. For each table, do the following:

a. Describe symbolically or in words the pattern hiding in the table.

b. Give the missing values in the table.

c. Tell whether the relationship between x and y in the table represents a quadratic, linear, or exponential relationship. Explain how you know.

1.

x	0	1	2	3	4	5
y	1	3	9	27		

2.

x	−1	0	1	2	3	4	5
y	−5	0	3	4	3		

Table Patterns II

Look at the tables below. For each table, do the following:

a. Describe symbolically or in words the pattern hiding in the table.

b. Give the missing values in the table.

c. Tell whether the relationship between x and y in the table represents a quadratic, linear, inverse, exponential, or other relationship. Explain how you know.

1.

x	−1	0	1	2	3	4	5
y		0	−4	−6	−6	−4	

2.

x	1	2	3	4	5
y	0.5	0	−0.5		

Answer Key

Part 1: Number and Quantity

1.
   1. $\frac{1}{x^n}$
   2. y^{r-s}
   3. x^{ab}
   4. $a^{[illegible]} b^{[illegible]}$
   5. $\frac{d^n}{t^n}$
   6. x^{-n}
   7. n^{p+q}
   8. $\sqrt[n]{y}$

2. Answers will vary. Sample answer: Not all quadratic functions have x-intercepts, but they do all have a y-intercept. Quadratic functions can have 2, 1, or 0 x-intercepts depending on the location of the vertex and the direction of the opening. One example: If the quadratic function has a vertex below the x-axis and opens down, then it has no x-intercepts.

3. Answers will vary. Sample answer: An advantage is that matrices are easily stored and calculated with technology. Disadvantages are that matrices only represent two variables in the rows and columns and are not as visual as graphs. A matrix could be represented in different matrices by transposing the rows and columns and by listing the data in a different order.

4. **Mountain bike sales**

	Trek	K2	Schwinn
Los Angeles	100	55	35
Boston	45	40	40
Chicago	30	60	25

Boston sales

	hybrid	mountain
Trek	30	45
K2	35	40
Schwinn	50	40

Combined sales all cities

	hybrid	mountain
Trek	135	175
K2	125	155
Schwinn	270	100

Students used addition and subtraction of matrices to construct the matrices.

Answer Key

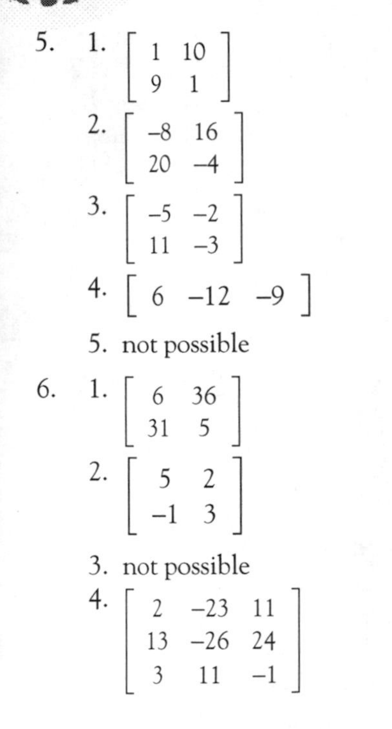

5. 1. $\begin{bmatrix} 1 & 10 \\ 9 & 1 \end{bmatrix}$

   2. $\begin{bmatrix} -8 & 16 \\ 20 & -4 \end{bmatrix}$

   3. $\begin{bmatrix} -5 & -2 \\ 11 & -3 \end{bmatrix}$

   4. $\begin{bmatrix} 6 & -12 & -9 \end{bmatrix}$

   5. not possible

6. 1. $\begin{bmatrix} 6 & 36 \\ 31 & 5 \end{bmatrix}$

   2. $\begin{bmatrix} 5 & 2 \\ -1 & 3 \end{bmatrix}$

   3. not possible

   4. $\begin{bmatrix} 2 & -23 & 11 \\ 13 & -26 & 24 \\ 3 & 11 & -1 \end{bmatrix}$

7. Answers will vary. Students should mention the importance of the dimension conditions. Yes, order matters: $A \times B \neq B \times A$.

8. Answers will vary. Sample answer: Matrices can be added if they have the same dimensions. They can be multiplied if the columns in matrix A match the rows in matrix B. The dimensions of the product matrix will be the number of rows in A × the number of columns in B. If the determinant of the matrix is 0, then it has no inverse.

   $A = \begin{bmatrix} 2 & 4 \\ 3 & 6 \end{bmatrix}$

   The inverse of matrix $A = \begin{bmatrix} 5 & 3 \\ 4 & 2 \end{bmatrix}$ is $A^{-1} = \begin{bmatrix} -1 & \frac{3}{2} \\ 2 & -\frac{5}{2} \end{bmatrix}$.

   The opposite of matrix A is $\left\{\begin{bmatrix} -5 & -3 \\ -4 & -2 \end{bmatrix}\right\}$. The identity matrix for these cases is $\begin{bmatrix} 1 & 0 \\ 0 & 1 \end{bmatrix}$.

Answer Key

Part 2: Algebra

9\.
1. $2\sqrt{3}$
2. $\frac{\sqrt{6}}{3}$
3. $\sqrt{5}$
4. $11\sqrt{2}$
5. $15\sqrt{3}$
6. $\frac{-13i}{13} = -i$
7. $\frac{(a+bi)^2}{a^2+b^2}$

10\.
1. correct
2. $5x$
3. $6n^2$
4. $6xy$
5. $a + b$
6. $n + 5/n$ or $1 + 5/n$
7. correct
8. $2x^3$
9. correct
10. $5a + 7b$

11\.
1. $x = 3.49$
2. $x = 1,194$
3. 9log2 + 3log5
4. 8logx
5. $x = -4$

12\.
1. $y = \frac{2}{x}$
2. $y = \frac{1}{2x}$
3. $y = 2x^3$
4. $y = 1$
5. $y = x^2$
6. $y = 3^5 = 243$
7. $y = 5^4 = 625$

13\.
1. $y = -16t(t + 3)$
2. $y = 5x(x - 2)$
3. $y = (3 + x)(x - 1)$
4. $y = 13(4x^2 - 1) = 13(2x - 1)(2x + 1)$
5. $y = (x + 4)(x + 1)$
6. $y = 2x^2 + 5x + x^2 - 10x = 3x^2 - 5x = x(3x - 5)$

14. 1. −1/2 and 3
    2. $x = -1$
    3. $\frac{-2 \pm \sqrt{-8}}{2}$, which means there is no intersection with the x-axis
    4. $\frac{-3 \pm \sqrt{17}}{4}$

    For all quadratic functions, the roots or zeros represent the intersection of the graph with the x-axis.

15. 1. $x + 2$
    2. not possible
    3. b
    4. $\frac{1}{f}$
    5. $x + 5$
    6. $n - 4$

16. $76 < (25 + w)(50 + w) - 25(50) \leq 400$; Solving the two resulting inequalities results in lengths for w that could be between 1 meter and 5 meters.

17. 1. Eric weighs about 104 pounds.
    2. If Amber and Aleah continue to sit at 4 feet, then $130 \times 4 = 104 \times D$, and $D = 5$ feet. If the seesaw is long enough, then Eric can balance them. Otherwise, he cannot.

18. $70m + 100 > 500$; $70m > 400$; $m > 5.7$. School started 6 months ago.

19. 1. speed = 1.25 kilometers per minute
    2. $d = 1.25t$
    3.

Time	20	40	60	80	100
Distance	24.5	50	75	100	125

    4. 24.5 kilometers
    5.

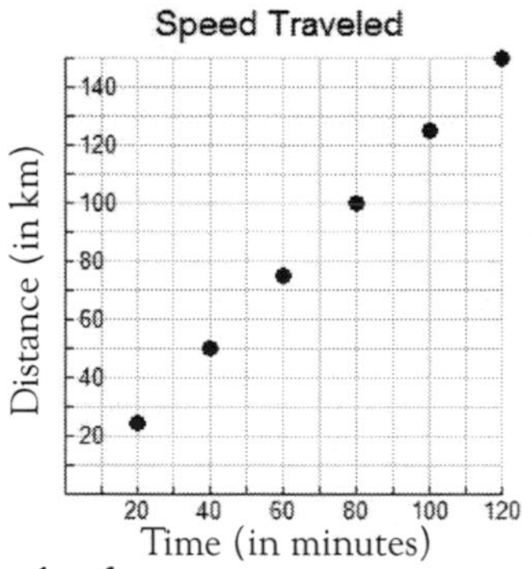

Answer Key

20. Points will vary.
    1. $y = |x|$
    2. $y = x^2$
    3. $y = \frac{1}{x}$
    4. $y = x^3$

21. Toppings and soft drink

    Soft drinks: 2 4 6 8 10 12 14 16 18 20

    Toppings: 3 6 9 12 15 18 21 24 27 30 33

    Corner points: (0, 0), (0, 16) and (32, 0). Boundaries: $.5t + 1.00s \le 16$ and $t \ge 0$, and $s \ge 0$

22. Students might talk about undoing the steps and subtracting 17 from (or adding −17 to) both sides and then dividing by −4.

23.
    1. 4
    2. 6
    3. 1
    4. 0, 3.5
    5. 0, 13
    6. 0, −2
    7. 0, 5
    8. −2
    9. −2
    10. 0, 7

24. Answers will vary. Sample answers:
    1. C might represent the original cost of a coat with a discounted price of $37.49. C = $49.99.
    2. The sale price of a shirt might be $28.00 after a 20% discount. C represents the original cost of the shirt. C = $35.00

25. Students might discuss how the equation $x^2 = -9$ cannot be solved in the set of real numbers because the square of any number is a nonnegative number. Students might also talk about the graph of the equation $y = x^2 + 9$ and show that it does not intersect the x-axis, so there are no "zeros" or roots for $y = 0$.

26. Original equation: $y = f(x) = x^2 + 8x + 10$
    Roots: $x = -4 \pm \sqrt{6}$; x-intercepts are $(-4-\sqrt{6},0)(-4+\sqrt{6},0)$.

27. Zeros: 3, −1; vertex: (1, −4); intercepts: (0, −3), (3, 0), (−1, 0). Other answers will vary.

28. $C = 20{,}000 + .02p$, $C = 17500 + .025p$; The school must make more than 500,000 copies before the higher price is a reasonable choice.

29.
    1. $x + y = 10$; $x + 2y = 8$; (12, −2)
    2. $x + 2y$; $x + y = 15$; (10, 5)
    3. $x - y = 2$; $2x - 2y = 4$; Any values for x and y that differ by 2 will work in this puzzle. There are infinite solutions.

30. Word puzzles will vary.
    1. (2, 1)
    2. (−2, 4)
    3. not possible

31. Students' choices and rationales may vary.
    1. (−4, −5)
    2. (−6, 4)
    3. (−20, 10)
    4. (3.6, 2.4)

32. Answers will vary.
    1. Sample answer: to find an ordered pair that satisfies both equations
    2. Sample answer: Graph both lines on the same set of axes, and locate the intersection point of the two lines.
    3. Sample answer: Substitute values of x into both equations. Look for identical y-values for the same value of x.
    4. Sample answer: solve algebraically by substitution or elimination
    5. Sample answer: If the values in the table have the same constant rate of change, if the lines are parallel, or if the slopes are the same but have different y-intercepts, then there is no solution.

33. The time is 2.02 seconds when the height is 10 meters. Students may use a graphing calculator and the trace feature to find the answer.

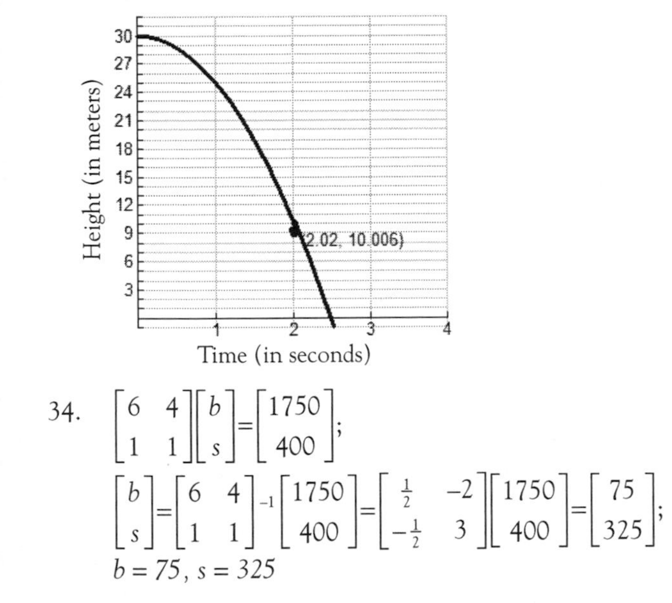

34. $\begin{bmatrix} 6 & 4 \\ 1 & 1 \end{bmatrix}\begin{bmatrix} b \\ s \end{bmatrix}=\begin{bmatrix} 1750 \\ 400 \end{bmatrix}$;

$$\begin{bmatrix} b \\ s \end{bmatrix}=\begin{bmatrix} 6 & 4 \\ 1 & 1 \end{bmatrix}^{-1}\begin{bmatrix} 1750 \\ 400 \end{bmatrix}=\begin{bmatrix} \frac{1}{2} & -2 \\ -\frac{1}{2} & 3 \end{bmatrix}\begin{bmatrix} 1750 \\ 400 \end{bmatrix}=\begin{bmatrix} 75 \\ 325 \end{bmatrix};$$

$b = 75$, $s = 325$

35.
1. A table of values would have increasing values of x and simultaneously increasing values of y. The graph would slope up and to the right.
2. This situation would have x values increasing and y values decreasing, so the slope of the graphed line would be down and to the right.
3. The points lie on a straight line when there is a common slope between each set of points.
4. The points could be connected if the data expressed represents a continuous change in values.

36.
1. $x + y \leq 75$ and $y \leq .5x$ and $y \geq 0$ and $x \geq 0$
2.

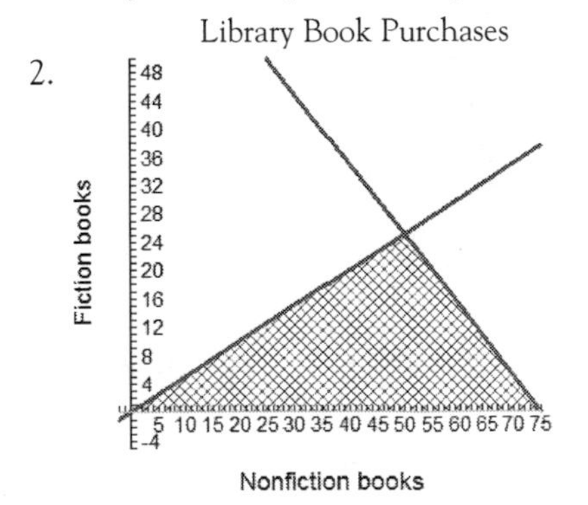

3. The intersection of the shading identifies the distribution of fiction and nonfiction books for her purchases.
4. Marian can buy 50 nonfiction and 25 fiction books.

37. 1.

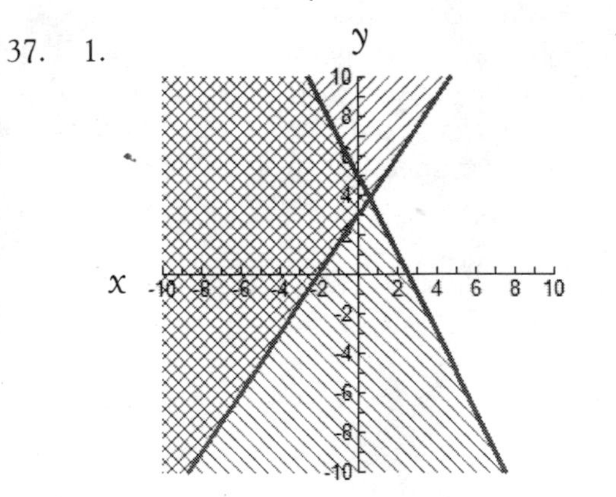

2.

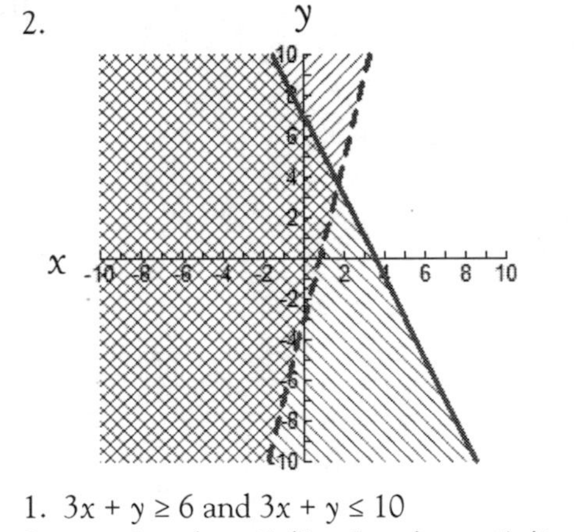

38. 1. $3x + y \geq 6$ and $3x + y \leq 10$
2. $x \geq -3$ and $y \leq 5x/2 + 5$ and $y \leq -5x/3 + 5$

Part 3: Functions

39. Tables *a* and *c* could be functions. Tables *b* and *d* could not. In table *b*, $x = 3$ is paired with two different values of y. In table *d*, $x = a$ is paired with two different values of *y*. This contradicts the definition that for any *x* there will be one and only one *y*.

40. 1. 115° C
    2. 36° C
    3. 20° C

41.

x	y
0	200
1	164
2	134.48
3	110.27
4	90.42
5	74.15
6	60.80

The 200 represents the measure of the starting quantity, and the 0.18 represents an 18% decay rate.

42. 1. $h(3) = -4.9(3)^2 + 150 = 105.9m$. This represents the height of the tennis ball 3 seconds after it is released.
    2. $-4.9t^2 + 150 \leq 25$, $t \geq 5.05$ sec
    3. $(-4.9t^2 + 150 = 0, t)$; $t = 5.53283$, $t = -5.53283$. In this problem, −5.53 has no meaning.

43. 1. $A = (n(120 - 2n)$ expresses the area of the pen when one side is against the barn, where n = width. $(n = 30; m = 60)$
    2. $n = 40$; $m = 40$; $A = 1{,}600$
    3. The maximum area using the barn is 1,800 square feet.
    4. Minimum values will vary depending on students' decisions on how short a width can be. One possible answer is 118 square feet using the barn.
    5. The maximum area without using the barn is 900 square feet. Minimum areas will vary; one possible answer is 59 square feet.

44. $I = (.20 + .02x)(10000–200x)$, where I is the income and x is the number weeks to wait; $x = 20$ weeks. The maximum occurs at the vertex.

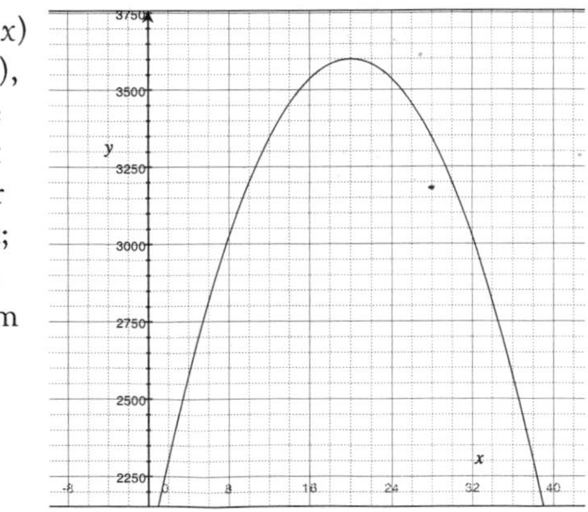

45. 1. arithmetic, linear, decreasing
    2. geometric, nonlinear, decreasing
    3. arithmetic, linear, increasing
    4. geometric, nonlinear, decreasing
    5. geometric, nonlinear, increasing

46. 1. $C = f(T) = T - 37$
    2.

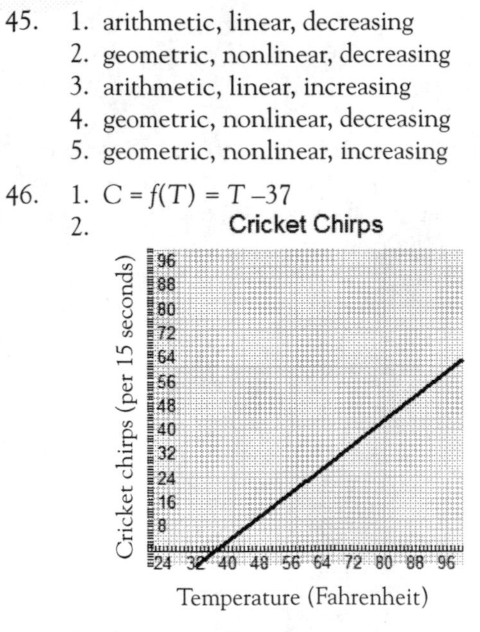

    3. Answers will vary.

47. 1. slope = −600/2 or −300
    2. rate of change = 300 mps downward
    3. The domain for the line segment is $1 \le s \le 3$. The range for the line segment is $1800 \le A \le 2400$.

48. 1.

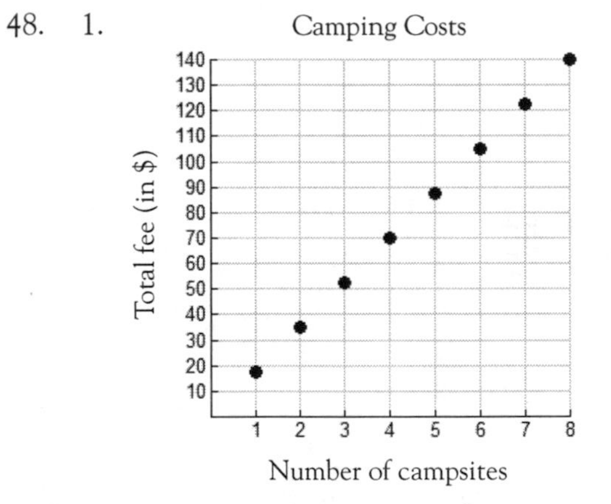

    It would not make sense to connect the points on the graph. The data points are discrete, not continuous.

    2. As the number of campsites increases, the fee increases a constant amount. The graph shows this with a slope up and to the right.

Answer Key

49. 1. 2.03 million per year from 1900 to 2000 is less than the 3.5 million per year from 2000 to 2010.
    2. Highest rate of change: 1950 to 1960 or 1990 to 2000. Lowest rate of change: 1930 to 1940. Calculate the rate of change for each 10-year period. Factors might include the postwar baby boom, the Great Depression, and so forth.
    3. Answers will vary. Sample answer: 348,000,000

50. 1. y-intercept: (0, 16); x-intercepts: (−4, 0), (4.3, 0). Other answers will vary.
    2. y is increasing over the interval (−5, 0).
    3. No, the rate of change is not constant. The change over (0, 2) is −1 , over (2, 4) is −7/2, and over (4, 5) is −7.

51. Answers will vary. Sample answer: For $a > 0$, would appear in quadrants I and III; for $a < 0$, it would appear in quadrants II and IV. For $a > 0$, $y = \frac{a}{x^2}$ would appear in quadrants I and II; if $a < 0$, it would appear in quadrants III and IV. Students might also discuss asymptotes and where the functions are increasing and decreasing.

52. Substitution into the given function or graphing and tracing yields results to the question. Graphing: $H(t) = -16t^2 + 120t + 6$; the highest point is 231 feet after 3.75 seconds. The ball will stay in the air for approximately 7.5 seconds.

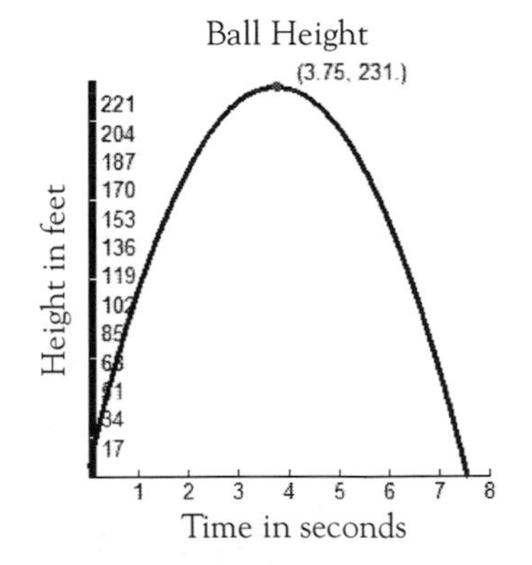

53. 1. Molly might be thinking about finding the break-even prices to charge. Lucas might be thinking about the fixed cost and about the change in sales if the ticket price is increased. Yes, the equations are equivalent.

2. Break-even prices are \$4 and \$16. Fixed cost is \$4,800. For every \$1 increase in ticket price, 75 fewer tickets would be sold. The best price might be \$10. Explanations will vary. Some students might use the trace feature on their calculator to determine the maximum values.

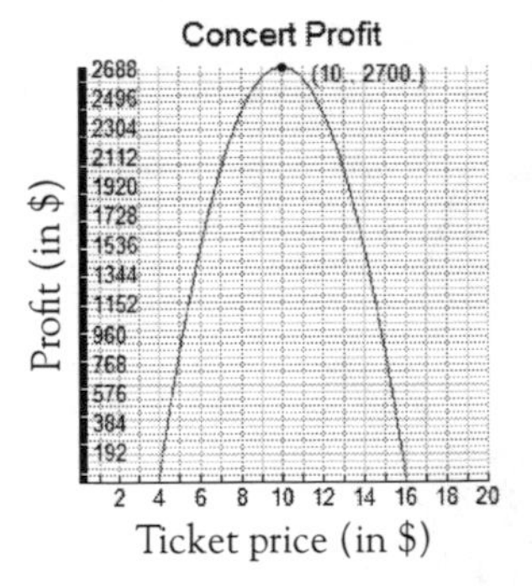

54. 1. Both are parabolic functions with vertex at (0, 0).

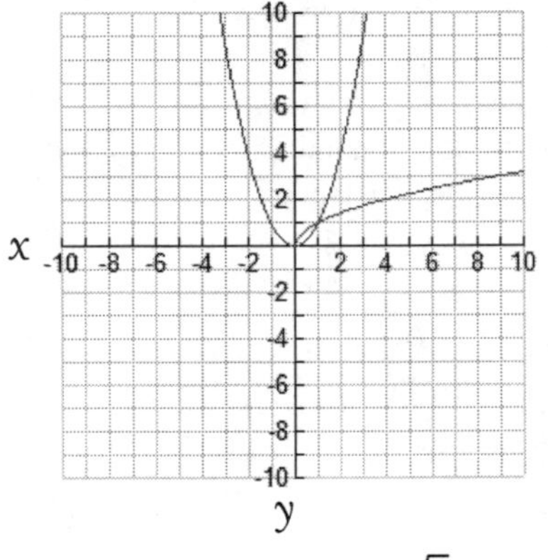

2. No, the graph of $y=\pm\sqrt{x}$ would not pass the vertical-line test and would not be a function.

55. Answers will vary. Sample answer: If $b = 0$ and $c = 0$, the graph of $y = x^2$ will stretch or shrink and will be symmetric to the y-axis. If $a < 0$, the graph will open downward. If $a > 0$, the graph will open upward. If $a = 1$ and $b = 0$, the value c represents the y-intercept, and the graph $y = x^2$ will be transposed along the y-axis.

Answer Key

56. 1. Penny's graph is a little off. The x-intercepts for this graph should be (6, 0) and (−2, 0). The vertex is at (2, −16) rather than (−2, −16).
    2. vertex form: $y = (x - 2)^2 - 16$; polynomial form: $y = x^2 - 4x - 12$

57. Equations 1 and 3 are parabolas with minimum points. In each, x is squared and the coefficient of x is positive. Equations 2 and 4 are also parabolas, but with maximum points; e is linear.

58. These are all parabolas or quadratic functions. Their rates of change switch from decreasing to increasing (or increasing to decreasing) at their vertex, which represents their maximum or minimum point. The coordinates of the vertices for these functions are (−3.5, −8.25) and (3.5, 16.25). The axis of symmetry passes through these vertices such that $x = -b/2a$. The y-intercepts are (0, 4) for the first two equations and (0, c) for the third equation.

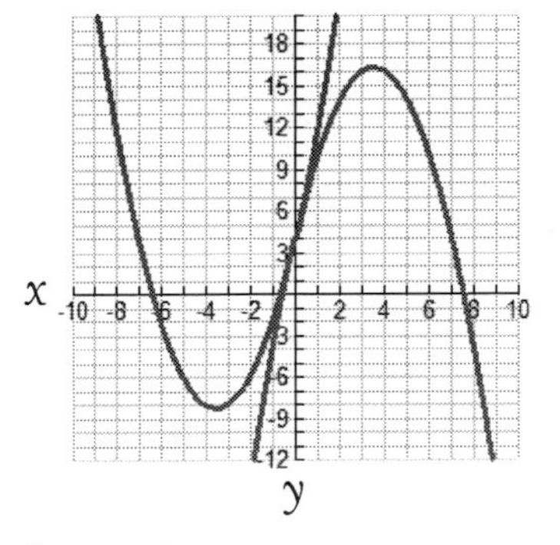

59. 1. no x-intercepts
    2. two x-intercepts
    3. one x-intercept
    4. two x-intercepts

    Students can determine answers using discriminants.

60. 1. $x \neq 0$, $y \neq 7$, vertical asymptote: $x = 0$, horizontal asymptote: $y = 7$
    2. $x \neq 5$, $y \neq 3$, vertical asymptote: $x = 5$, horizontal asymptote: $y = 3$
    3. $x \neq 0$, $y \neq 2$, vertical asymptote: $x = 0$, horizontal asymptote: $y = 2$

61. Answers will vary. The graphs of *a* and *c* lie completely in the first and third quadrants, and the graphs of *b* and *d* lie completely in the first and second quadrants. All approach zero as *x* approaches infinity. Asymptotes are $x = 0$ and $y = 0$ for all graphs. All are functions.

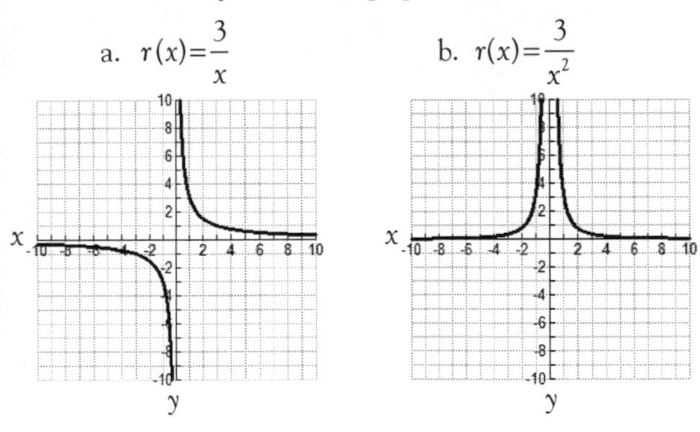

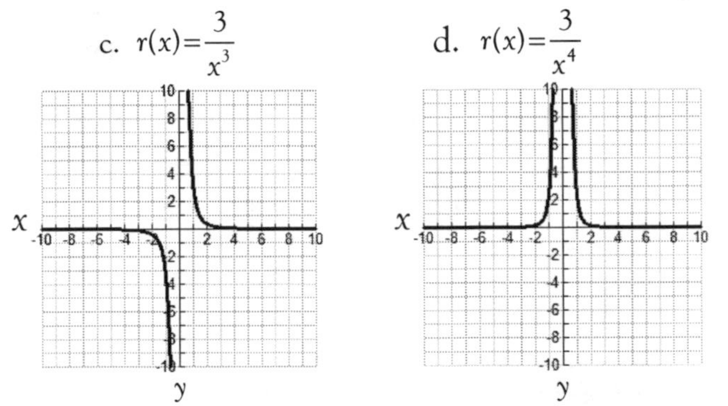

62. 1. As *x* gets very large positively, the functions increase rapidly. The *y*-intercept is $a + c$.
 2. When *a* is negative, the graphs will reflect over the *x*-axis. Changing *b* to $1/b$ reflects the graph over the *y*-axis.

63. Rita is confused about the definition of a function. While it is true that the line $y = 5$ intersects the graph of each equation in two places, this indicates that for any *y* value there may be two values of *x*. Both equations are functions because a vertical line cannot intersect these graphs in any more than one point.

Thus for any value of x, there is one and only one value of y.

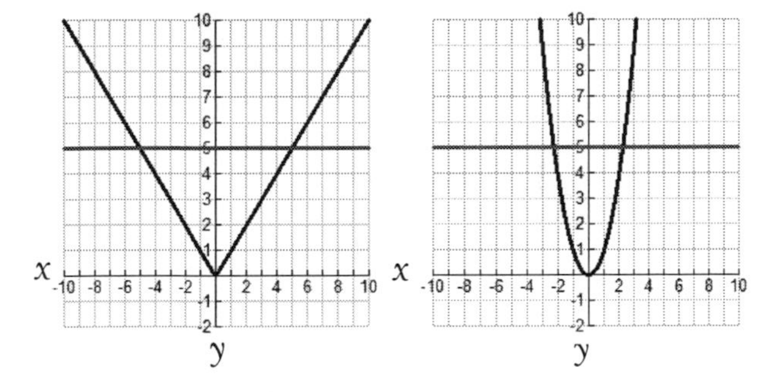

64. The functions represent exponential functions. For equations of the form $h(x) = a^x$ and $a > 1$, $h(x)$ is increasing. For $0 < a < 1$, $h(x)$ is decreasing. An asymptote exists at $y = 0$. The graphs have a y-intercept at (0, 1). There are no maximum or minimum values, and there are no symmetry features.

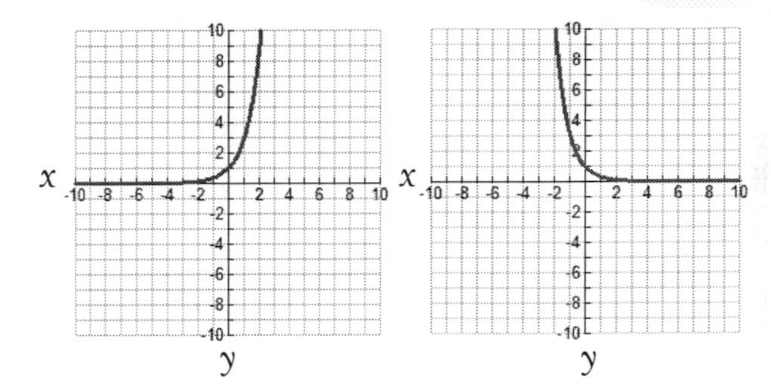

65. The function is called a *step function* because that's what it looks like on the graph. It is also known as the greatest integer function because the value of x is equal to the smallest integer not greater than x in the sentence $y = [x]$.

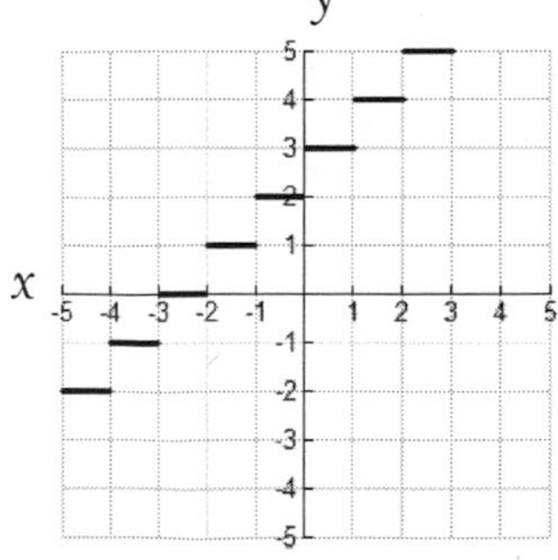

66. Answers will vary. A discussion of rational functions is likely. Some equations that would create a hole or an asymptote at $x = 2$ are
$y = \frac{(x-2)^2}{x-2}, y = \frac{1}{x-2}, y = \frac{1}{(x-2)^2},$ and $y = \frac{x-2}{x-2}$.

67. 1. For very large or very small values of x, the function approximates the graph of $g(x) = 2x + 17$. $f(x)$ can be rewritten as $g(x) + h(x)$. Carol's viewing window looks like $g(x)$.
    2. Each friend has a different window for the graph of the function.
    3. The function is undefined at $x = 2$, and there is an asymptote for the graph at that vertical line.

68. Juan is correct. These situations can be represented as inverse functions.

Tables:

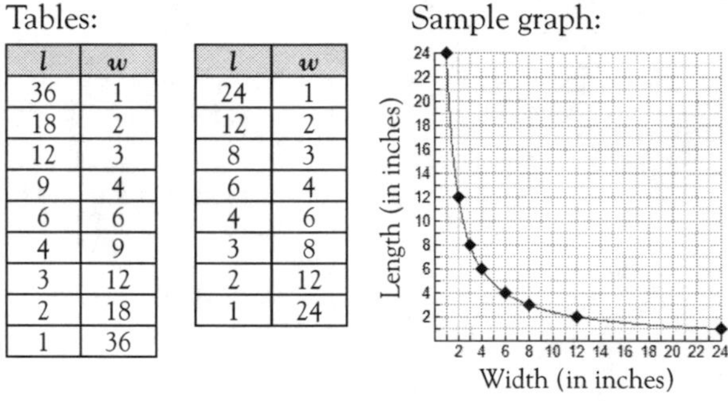

l	w
36	1
18	2
12	3
9	4
6	6
4	9
3	12
2	18
1	36

l	w
24	1
12	2
8	3
6	4
4	6
3	8
2	12
1	24

Sample graph:

Possible equations: $lw = 24$, thus $y = \frac{24}{x}$ could be a representation of this function for $x = w$ and $0 < w \le 24$.

69. 1. \$3.10
    2. \$2.86
    3. $y = 2.98(1+.04)^x$
    4. In year 5, it will surpass \$3.50.

70. $A = LW$ and $L + 2W = 1200$, $\frac{L + 2W}{2} \ge \sqrt{2LW}$, $600 \ge \sqrt{2A}$, $A \le 180{,}000$

71.

x	y
–2	1,600
–1	320
0	64
1	12.8
2	2.56
3	.512
4	.1024

A graph of the function would look as follows. This represents an exponential decay model:

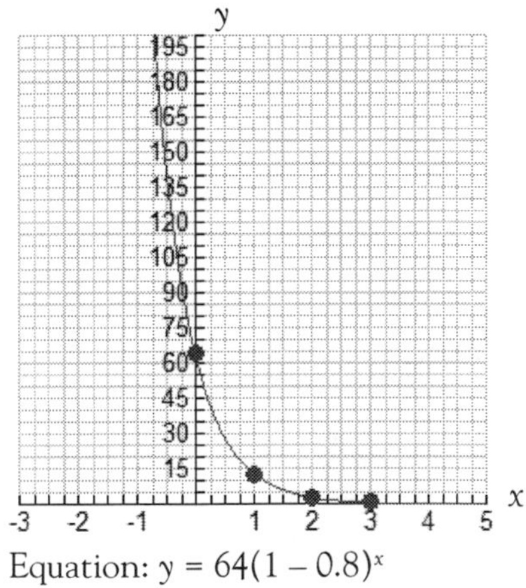

Equation: $y = 64(1 - 0.8)^x$

72. Answers will vary. A graph of the combined data is given. An argument could be made for either plan. If Rosa's family rents 25 DVDs per month, but they only watch 2 movies per month on average, the Deluxe company is a better choice for her family.

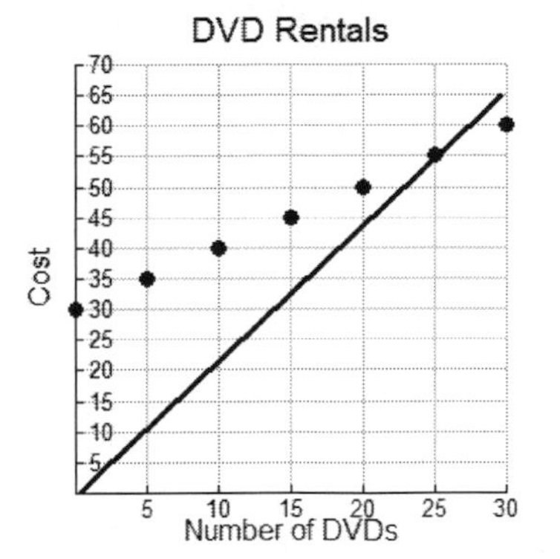

73.

Side of square cutout(s)	Length of box $(12 - 2s)$	Width of box $(9 - 2s)$	Volume of box $s(12 - 2s)(9 - 2s)$	Surface area of box
1	10	7	70 in³	104 in²
2	8	5	80 in³	92 in²
3	6	3	54 in³	72 in²
4	4	1	16 in³	44 in²

Greatest volume results from 2-inch cutouts. Smallest volume is from 4-inch cutouts.

74. The unit cube has 8 corners, 6 faces, and 12 edges.
$T = n^3$

L1	L2
1	1
2	8
3	27
4	64
5	125
6	256
7	343
8	512
9	729
10	1000

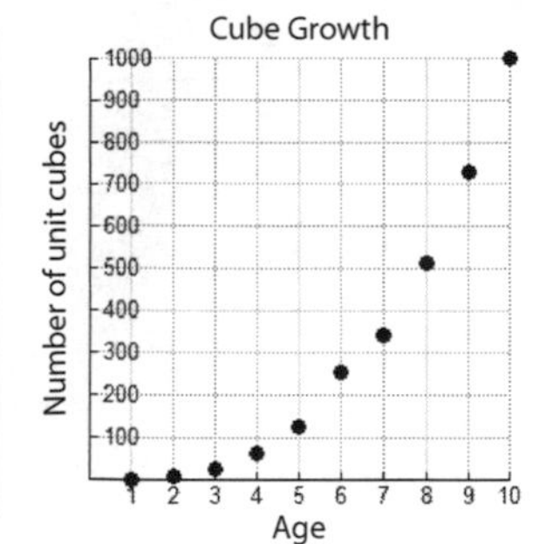

75. 1. 15 moves

2.

Number of each color	1	2	3	4	5	6	7	8	9	10
Number of moves	3	8	15	24	35	48	63	80	99	120

3. $M = n(n + 2)$

76.

Original square		Rectangular plot			Difference in areas
Side length	Area	Length	Width	Area	
4	16	7	1	7	9
5	25	8	2	16	9
6	36	9	3	27	9
7	49	10	4	40	9
8	64	11	5	55	9

This is not a fair trade with respect to area. The rectangle will always be 9 square meters smaller than the square. The area of a square = n^2. The area of a rectangle $=(n+3)(n-3) = n^2 - 9$.

77. 1.

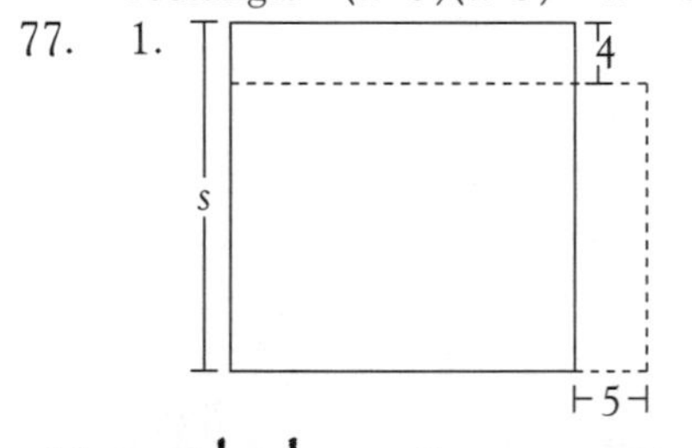

Answer Key

2. area of square = s^2, area of rectangle = $(s+5)(s-4)$

s	Area of square	Area of rectangle
5	25	10
10	100	90
15	225	220
20	400	400
25	625	630
30	900	910
35	1,225	1,240

3. For $s > 20$, the area of the rectangle will be greater than the area of the square.
4. For $s < 20$, the area of the rectangle will be less than the area of the square.
5. For $s = 20$, the areas will be equal.
6. Answers will vary.

78. 1. $C = 4.75(n) + 1.00$, C = total cost, n = number of packages

2. 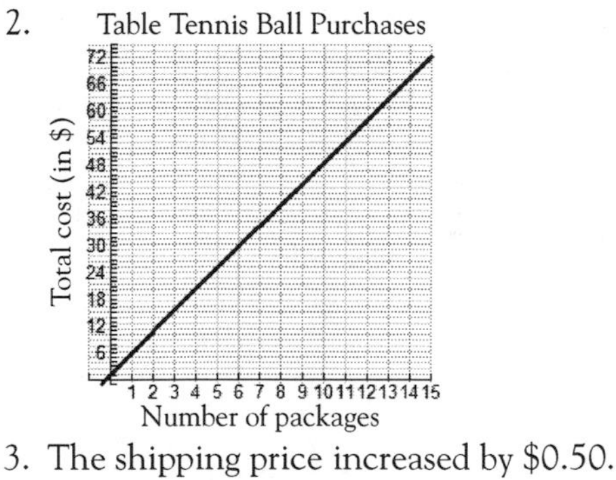

3. The shipping price increased by $0.50.
4. $C = 4.75(n) + 1.50$

79.

Number of toothpicks	6	11	16
Number of hexagons	1	2	3
Perimeter of figure	6	10	14

$T = 5H + 1$; $P = 4H + 2$

80. The fifth figure will have 15 dots. The sixth figure will have 21 dots. The equation is $T = \frac{n(n+1)}{2}$, where T is the value of the triangular number and n is the number of the term.

81. 1. This is a representation of square numbers.
    2. 100 blocks
    3. 10,000 blocks
    4. $B = n^2$

82. You might want to encourage students to think about square numbers and triangular numbers.
    $P_n = S_n + T_n = n^2 + \frac{n(n-1)}{2} = \frac{n(3n-1)}{2}$.

83. 1. 40 pages
    2. 20 and 21; 41
    3. One possibility: N sheets of paper result in $4N$ pages. The sum of the page numbers on one side of 1 sheet is $4N + 1$.
    4. 820
    5. $2N$ represents the number of sides of each sheet. Thus, the sum of the pages $= 2N(4N + 1)$.

84. One way to think about this situation is to imagine the stacked boxes are a staircase that is half of a rectangle, where n represents the base and $n + 1$ represents the height of the rectangle. The value $2B$ then would be equal to $n(n + 1)$. In other words, twice the number of boxes must equal the product of two consecutive integers. Solving for the equation $2B = n(n + 1)$ where $B = 45$ gives a quadratic where $n = -10$ or $n = 9$. Since Freya can't use a negative number of boxes, she should use 9 boxes on the bottom row.

85. Paolo could make 29 pieces with 7 cuts.

Number of cuts	0	1	2	3	4	5	6	7	8	9	n
Number of pieces	1	2	4	7	11	16	22	29	37	46	$\frac{n(n+1)}{2}+1$

86. 1. 625; no middle number in row 50; n^2
    2. 6; 8; 78; $2(n - 1)$
    3. 29; 41; $n^2 + n - 1$
    4. n^3
    5. $\left[\frac{n(n+1)}{2}\right]^2$

87. $D = 60t$. Letters will vary.

Answer Key

88. Answers will vary. Sample answers: $4(S + 1)$; $2S + 2(S + 2)$; $(S + 2)^2 - S^2$; $4(S + 2) - 4$; $S + S + S + S + 4$

89.

U-Say	3	0	–4	1	2	5
I-Say	11	2	–10	5	8	17

Abby's rule: U = 3 • I + 2

90. 1.

Side length	3 faces	2 faces	1 face	Zero faces
2	8	0	0	0
3	8	12	6	1
4	8	24	24	8
5	8	36	54	27
6	8	48	96	64
7	8	60	150	125
8	8	72	216	216
9	8	84	294	343
10	8	96	384	512
15	8	156	1014	2197

2. 3 faces located in corners; 2 faces located on edges; 1 face located on faces other than those listed above; 0 faces located on the interior
3. For all cubes, 3 faces is 8; 2 faces: $N = 12(s - 2)$; 1 face: $N = 6(s - 2)^2$; zero faces: $N = (s - 2)^3$.

91. 1. $2,030.00
2. $2,060.45
3. $2,090.90
4. After 15 months, his unpaid balance will be greater than $2,500.

92. Car B will be worth more than Car A after 3 years.

Year	Car A	Car B
0	15,000	12,000
1	15,000 × .70 = 10,500	12,000 × .80 = 9,600
2	7,350	7,680
3	5,145	6,144

93. $B(x) = 200(1 - .25)^x$. Bounce 5 will be less than 50 centimeters.

Bounce	Height
0	200
1	150
2	112.5
3	84.375
4	63.281
5	47.46
6	35.6
7	26.7
8	20
9	15
10	11
11	8

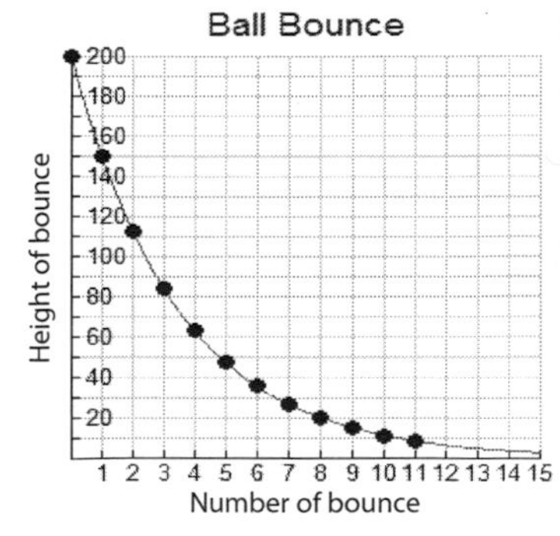

94. 1. a line of best fit through the points
    2. a constant rate of change for the ordered pairs
    3. first-degree values for x and y
    4. Answers will vary.

95. 1. quadratic
    2. linear
    3. quadratic
    4. exponential
    5. quadratic

96. 33 people provide the greatest income for EVA.
    $f(x) = 200 - 4(x - 10) = 240 - 4x$

97. $C_1 = 20.00 + 10m$; $C_2 = .40m$

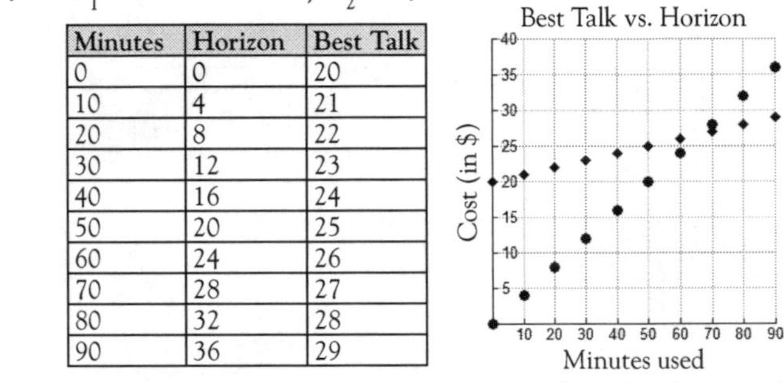

Minutes	Horizon	Best Talk
0	0	20
10	4	21
20	8	22
30	12	23
40	16	24
50	20	25
60	24	26
70	28	27
80	32	28
90	36	29

In the graph, Best Talk is represented by the diamond shapes and Horizon is represented by the circles. Students may argue that Horizon is the best plan until Aisha exceeds 65 minutes in a month.

98. 1. $T = 2000 + 399m$; $T = 1600 + 420m$
    2. At the end of 1 year, Joe's car will have cost $6,788 and Hank's car will have cost $6,640. At the end of 2 years, Joe's car will have cost $11,576, and Hank's car will have cost $11,680.
    3. Answers will vary. Students could make an argument for either plan.

Answer Key

99. 1. $y = -3x + 11$
    2. $y = -\frac{2}{3}x + \frac{5}{3}$
    3. $y = -\frac{5}{4}x + \frac{17}{2}$
    4. $y = -x + 1$
    5. $y = x + 0$

100. 1. $y = \frac{3}{2}x + 2$
    2. $y = 3$
    3. The line passes through (1, 5) and (4, 6), thus $y = \frac{1}{3}x + \frac{14}{3}$.
    4. $y = -x$

101. 1. $y = 2x + 3$
    2. $y = -3x + 5$
    3. $f(x) = \frac{1}{3}x - \frac{13}{3}$.
    4. $f(x) = \frac{2}{3}x + 5$

102. 1. $y = -300x + 350$ for x = cost and y = number sold
    2. The slope = −300, and the y-intercept = 350. For every $1.00 increase in price, the number sold decreases by 300; in addition, they can give away 350 brownies if they don't charge anything.
    3. 140 brownies
    4. about $0.17 for each brownie

103. 1–3. Answers will vary.

104. Answers will vary. Sample answer: If you know the slope and y-intercept, you can substitute the values into the equation $y = mx + b$ or $y = a + bx$. The slope can be determined from the graph by finding the change in y divided by the change in x. The y-intercept will be the point where $x = 0$ in the table and where the line crosses the y-axis on the graph.

105. Slope is $\frac{212-32}{100-0} = \frac{180}{100} = \frac{9}{5}$, thus $y = \frac{9}{5}x + 32$.
    Note: y = Fahrenheit and x = Celsius.

106.

Number of subscriptions sold	Weekly pay (dollars)
1	145
2	170
3	195
6	270
10	370
12	420
15	495

2. $W = 25S = 120$; $S = (W - 120)/25$

3. 20 subscriptions

107. 1. $C(d) = 1{,}200(2^d)$, Count $= 1{,}200(2^d) = 1{,}200(2^{3.25})$ $= 11{,}416$

2. A trace can be used to find the count at $x = 3.25$ for the number of days indicated in the question and for when the bacteria count goes over 200,000.

$200{,}000 = 1{,}200(2^d)$

$2^d = 166.66$

$d \approx 7.4$

Thus, the pool will need treatment on day 7.

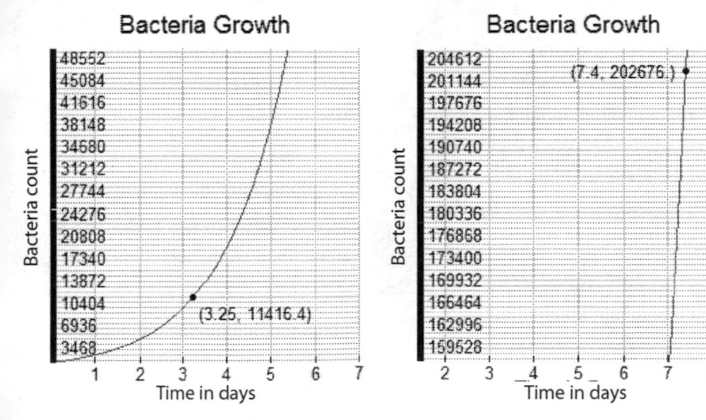

108. Answers will vary. Sample answer: The rate of change is the value found when the difference in two y values is divided by the difference in the corresponding two x values, either from the table or from the graph using any two points. In the equation, the rate of change is the value m or b when the equation is written in the form $y = mx + b$ or $y = a + bx$. The y-intercept is the value $(0, x)$.

109. Students could solve the equation substituting 12 for h, or they might draw a graph on their graphing calculators and estimate the intersection of two lines. For example, the graph below indicates that the height will be 12 feet at approximately 7 A.M. and again at 11 A.M. Melinda's father is too late by 1 hour in his prediction. The height will only be about 9.5 feet.

Tidal Change:

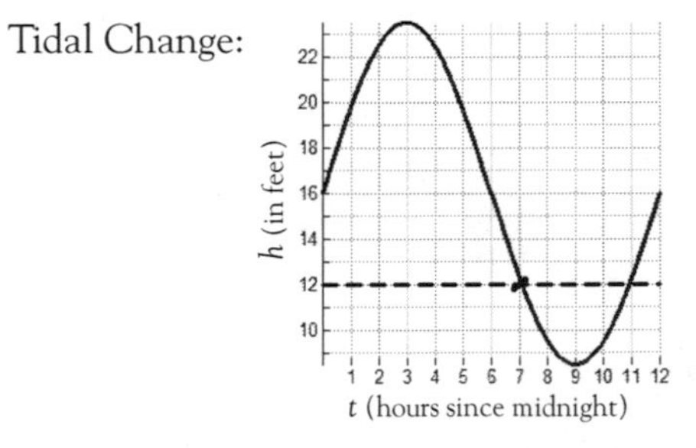

Answer Key

Part 4: Statistics and Probability

110. Students may need to be reminded to convert 12.5 feet to 381 centimeters. Students' equations may vary depending on the strategy used. This data and graph leads to an answer of 23 rubber bands.

111. 1.

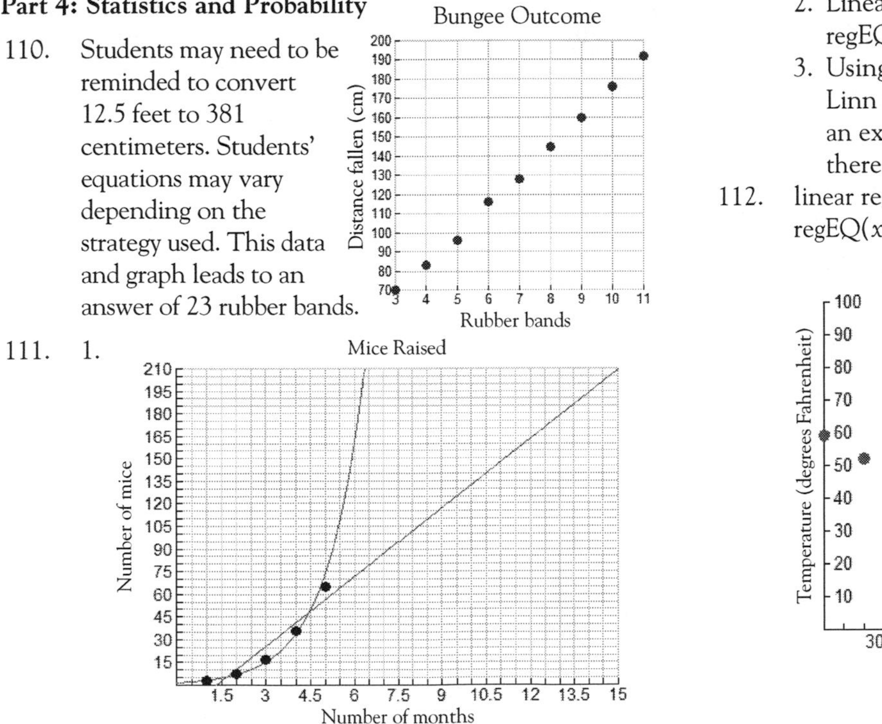

2. Linear regression $(a + bx)$
   $\text{regEQ}(x) = -20.3 + 15.3x$
3. Using a linear model as a predictor indicates that Linn might have 209 mice after 15 months. Using an exponential graph as a predictor indicates that there may be 200 mice by 7 months.

112. linear regression $(a + bx)$
$\text{regEQ}(x) = 59.0067 + -.003567x$

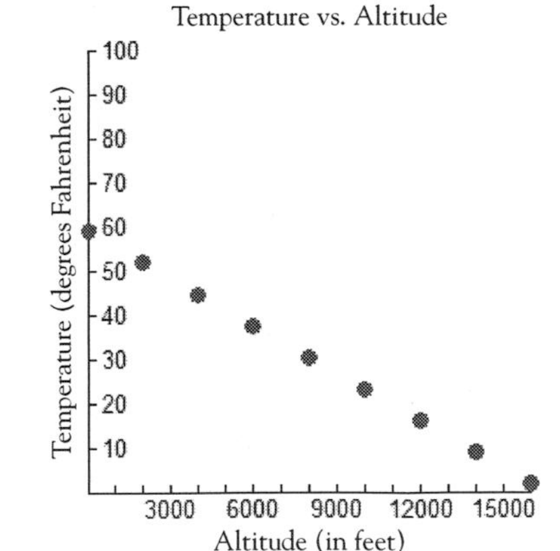

113. Solutions will vary depending on students' choices. One graph and linear regression equation are provided for the data for both sexes.

Linear regression $(ax + b)$

regEQ$(x) = .229738x + -381.817$

$a = .229738$

$b = -381.817$

$r = .981234$

$r^2 = .96282$

For this graph and equation, a student might predict that a person born in 2020 would have a life expectancy of 82.26 or 82.3 years.

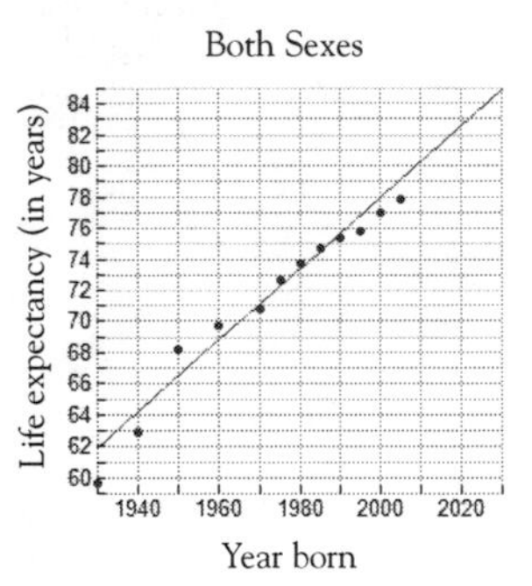

114. The weight increases more each month in the first year than in the second year. The graph looks like it might fit a parabolic curve for $0 \le x \le 25$.

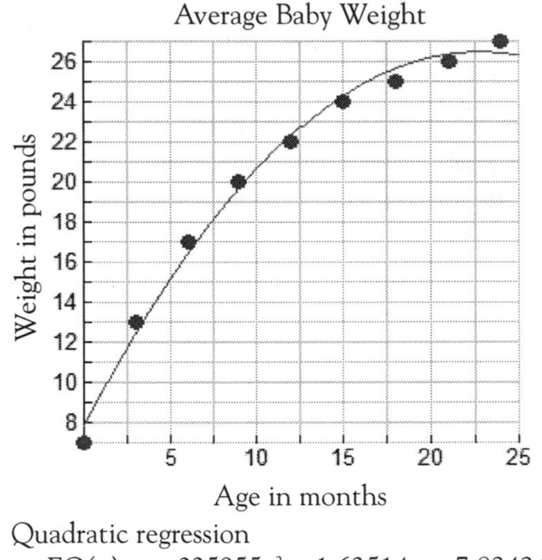

Quadratic regression

regEQ$(x) = -.035955x^2 + 1.63514x + 7.82424$

115. At 80 miles per hour, the stopping distance would be close to 500 feet. Explanations will vary.

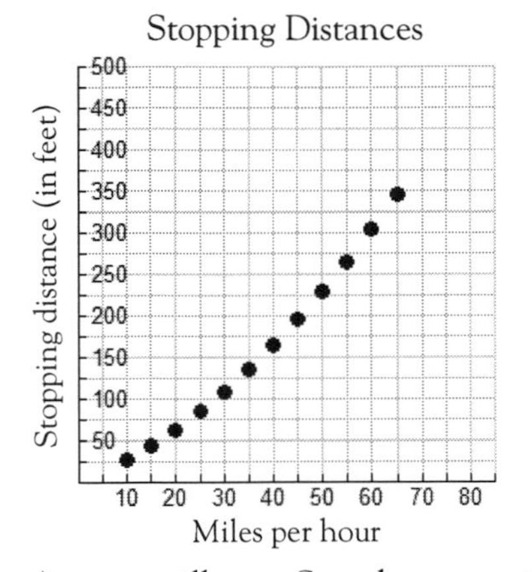

116. Answers will vary. Sample answer: One way to compare two linear equation models is to compare their average mean errors by calculating the difference between the predicted value and the actual value for various x values.

117. 1. linear regression $(a + bx)$
regEQ(x) = 8.1875 + .458333x
2–3. Answers will vary.

118. Answers will vary.

119. 1. a. $y = 3^x$; As x increases, y increases by a factor of 3. There is a constant growth factor.
b. missing values: (4, 81) (5, 243)
c. The relationship is exponential.
2. a. $y = -x^2 + 4x$; As x increases, y increases by a decreasing amount. Second differences are constant.
b. missing values: (4, 0) (5, −5)
c. The relationship is quadratic.

120. 1. a. $y = x^2 - 5x$; As x increases, y decreases by a decreasing amount. Second differences are constant.
b. missing values: (−1, 6) (5, 0)
c. The relationship is quadratic, since the second differences are constant.
2. a. $xy = 1/2$; As x increases, y is $1/2x$, since the second differences are constant.
b. missing values: (4, 1/8) (5, 1/10)
c. The relationship is inverse, since as x increases, $y = 1/2x$.